DEFECT ORIENTED TESTING FOR CMOS ANALOG AND DIGITAL CIRCUITS

FRONTIERS IN ELECTRONIC TESTING

Consulting Editor
Vishwani D. Agrawal

Books in the series:

DEFECT ORIENTED TESTING FOR CMOS ANALOG AND DIGITAL CIRCUITS

by

MANOJ SACHDEV
Philips Research

KLUWER ACADEMIC PUBLISHERS
BOSTON / DORDRECHT / LONDON

A C.I.P. Catalogue record for this book is available from the Library of Congress.

\ISBN 0-7923-8083-5

Published by Kluwer Academic Publishers,
P.O. Box 17, 3300 AA Dordrecht, The Netherlands.

Sold and distributed in the U.S.A. and Canada
by Kluwer Academic Publishers,
101 Philip Drive, Norwell, MA 02061, U.S.A.

In all other countries, sold and distributed
by Kluwer Academic Publishers,
P.O. Box 322, 3300 AH Dordrecht, The Netherlands.

Printed on acid-free paper

Printed in the Netherlands

This book is dedicated to my wife, Sunanda, and to our children, Aniruddh and Arushi.

Table of Contents

Foreword

We have made great strides in designing complex VLSI circuits. A laborious design verification process ensures their functional correctness. If no defects occur in manufacturing then testing will not be required. However, the world is not so perfect. We must test to obtain a perfect product.

An exact repetition of the verification process during manufacture is too expensive and even impossible. So, we test for a selected set of modeled faults. There is no unified modeling procedure for the variety of VLSI chips we make. Stuck-at model applies only to some types of digital circuits. Besides, there are problems, such as, (a) some stuck-at faults cannot occur in the given VLSI technology and (b) some actual manufacturing defects have no stuck-at representation. Numerous known problems with the present-day test procedures point to a defect-oriented testing. This simply means that we use the knowledge about the manufacturing process to derive tests. Such tests provide the greatest improvement in the product quality for the minimum cost of testing.

Dr. Sachdev has done original work on defect-oriented testing. He takes experimental defect data and applies the inductive fault analysis to obtain specific faults for which tests should be derived. His work is done in an industrial setting and has been put to practice at Philips Semiconductors and elsewhere. The material in this book is collected from his PhD dissertation, research papers and company reports.

A strength of this book is its breadth. Types of designs considered include analog and digital circuits, programmable logic arrays, and memories. Having a fault model does not automatically provide a test. Sometimes, design for testability hardware is necessary. Many design for testability ideas, supported by experimental evidence, are included.

In addition to using the functional and other conventional tests, Dr. Sachdev takes full advantage of the defect-isolating characteristics of non-functional tests. Imagine taking a multiple-choice examination. All of us can remember making a guess some time and succeeding. Suppose, I connect you to a lie detector while you checked those choices. The lie detector may tell me to fail you even on some correct answers. Also, given the new procedure, we can design special tests. Current measurements similarly bring out the *internal conflicts* whose effects may not be visible by conventional logic tests. Such tests, though non-functional, improve the defect coverage. Current measurement is an important subject discussed in this book.

Non-functional tests are not without their pitfalls. Not much is accomplished if one who is going to be an electrical engineer passes or fails an examination in history. Clearly, there is need for matching the test with the function. In electronic circuits a non-functional test, designed to isolate a real defect, can reject a circuit with some other functionally acceptable defect. This phenomenon, known as yield loss due to non-functional tests, impacts costs similar to the design for testability overhead. In both cases, the costs are associated with quality improvement. A central theme in this book is to minimize such costs and it wonderfully succeeds in putting the economics of test and manufacture into practice.

Vishwani D. Agrawal
Bell Labs
Murray Hill, New Jersey
va@research@bell-labs.com
September 1997

Preface

Defect oriented testing methods have come a long way from a mere interesting academic exercise to a hard industrial reality. Many factors have contributed to its industrial acceptance. Traditional approaches of testing modern integrated circuits (ICs) have been found to be inadequate in terms of quality and economics of test. In globally competitive semiconductor market place, overall product quality and economics have become very important objectives. In addition, electronic systems are becoming increasingly complex and demand components of highest possible quality. Testing, in general and, defect oriented testing, in particular, help in realizing these objectives.

The motivation for this book comes from two counts. Firstly, the field of defect oriented testing is nearly two decades old. However, the information is fragmented and distributed in various conference and journal papers. There is hardly any book providing a cohesive treatment this field deserves. In this book an attempt is made to bridge this gap and provide an overview of this field. The target audience of this book consists of design and test professionals. Secondly, there is a wide disparity among various companies as well as academic institutions on the knowledge of this subject. A vast majority of research is carried out by few companies and academic institutions. Therefore, it is intended that this book will help in spreading the knowledge of the subject.

Defect oriented testing is expected to play a significant role in coming generations of technology. Smaller feature sizes and larger die sizes will make ICs more sensitive to defects that can not be modeled by traditional fault modeling approaches. Furthermore, with increased level of integration, an IC may contain diverse building blocks. Such blocks include, digital logic, PLAs, volatile and non-volatile memories, analog interfaces, etc. For such diverse building blocks, traditional fault modeling and test approaches will become increasingly inadequate.

Acknowledgments

During the course of writing, I had occasions to discuss the subject with several individuals. I express my gratitude to all those who have contributed with their ideas, time, and effort towards the successful completion of this book. It is not possible to name all these individuals in a limited space. However, I would like to name some of the individuals particularly. I am thankful to Eric van Utteren, Cees Niessen, the management of Research Labs, for their moral support. I am indebted to Keith Baker of Philips Research Labs, Joan Figueras of Polytechnical University of Catalunya, Chuck Hawkins of University of New Mexico and Wojciech Maly of Carnegie Mellon University for various discussions on defect modeling, IC testing, and I_{DDQ} testing. I also acknowledge discussions with Bert Atzema, Eric Bruls, Taco Zwemstra, Rene Segers, and others, on defect oriented testing at Philips Research Labs. Furthermore, discussions with Tony Ambler of University of Texas at Austin on test economics were useful. Finally, the contribution of Vishwani Agrawal of Bell Labs for his painstaking efforts on the manuscript and on various discussions on test is gratefully acknowledged.

Manoj Sachdev
Philips Research Labs
Eindhoven, The Netherlands

CHAPTER 1

Introduction

This chapter introduces some key test issues, namely test complexity, quality, reliability and economics, faced by semiconductor test industry. These issues form a basis for subsequent chapters.

1.1 The Test Complexity

Imperfections in the manufacturing process necessitate the testing of the manufactured integrated circuits (ICs). The fundamental objective of the testing is to distinguish between good and faulty ICs. This objective can be achieved in several ways. Earlier, when the ICs were relatively less complex, this objective was achieved by functional testing. Functional tests are closely associated with the IC function. Therefore, these tests are comparatively simple and straightforward. A 4-bit binary counter can be exhaustively tested by $2^4=16$ test vectors. However, as the complexity of the fabricated ICs increased, it was soon discovered that the application of a functional test is rather expensive on test resources and is inefficient in catching the manufacturing process imperfections (or defects as they are popularly known). For example, a digital IC with 32 inputs has only a modest design complexity by today's VLSI standard, but will require $2^{32}=4,294,967,296$ test vectors for exhaustive functional testing. If these are applied at the rate of 10^6 vectors per second, it will take 71.58 minutes to test a single IC. The test becomes even longer if the IC contains sequential logic. Obviously, it is too expensive a test solution to be practical.

The test problem is further compounded by the rapid development of CAD tools in the areas of IC design and manufacturing which help engineers to design and fabricate complex ICs. The need for simulation tools for test and testability analysis became visible only when testing was recognized as a bottleneck in achieving increasingly important quality, reliability and time to market goals. The packaging limitation puts severe additional constraints on the test of complex ICs. For example, the packaging capabilities over the last decade have increased only by an order of magnitude. The state of the art today is a 500-600 pin package compared to a 50-60 pin package nearly a decade ago. On the other hand, in the same period, the device integration on an IC has increased by an order of three or more. Effectively, the depth of logic is to be accessed from primary pins increased for each successive generation of chips. In other words, controllability and observability objectives became much more difficult to achieve for modern ICs. As a result, test vector sequences became longer and added to the test cost. During the same period, the cost of a general purpose automatic test equipment (ATE) also increased significantly. A state of the art ATE can now cost a few million dollars. The expensive ATE and the longer test vector sequences pushed the test costs to unacceptable levels.

The above mentioned scenario matches perfectly with the evolution of semiconductor memory market. Each successive DRAM generation has grown in complexity by a factor of four and the access time has decreased by a factor 0.8 for each new generation. Therefore, testing time is increased by 3.2 times for each new generation. This results in tremendous increase in testing cost which prevents the cost per bit from coming down despite increased integration. For example, a recent study of DRAMs identifies the test cost, along with process complexity, die size and equipment costs as a major component in the future DRAM chip costs [15]. The test cost of a 64 Mbit DRAM is projected to be 240 times that of 1 Mbit DRAM. For a 64 Mbit DRAM, the test cost to total product cost ratio is expected to be 39%. If the conventional test methods are used, test cost will grow at a rapid rate. The SRAM test cost is also expected to follow a similar trend. Moreover, RAMs are the densest and one of the biggest chips ever to be tested. DRAM chip size has grown by 40 to 50% while the cell size has decreased by 60 to 65% for each successive generation. The chip size of 64 Mbit DRAM is expected to be in the range of 200 mm^2. Small feature size and huge chip size result in an enormous critical area [9] for defects. Since RAMs must be mass produced, their test strategies are under severe pressure to ensure the quality of the tested devices while maintaining the economics of the production. In other words, testing of RAMs in an efficient, reliable and cost effective manner is becoming an increasingly challenging task [17].

This IC test cost explosion is not limited to RAMs. Complex microprocessors, and mixed signal devices such as single chip television, are other representative examples. In fact, for many VLSI ICs, testing costs are approaching 50% of the total production cost [26]. For example, the test development time for complex single chip television IC manufactured by Philips is reported to be many man years!! Such developments have caused a surge of interest in economics of test [3]. A number of studies have been reported on test economics [1,7,32]. Dislis et al. [7] demonstrated that economic analysis can be a powerful aid in the selection of an optimal set of design for test strategies, and in the organization of production test process.

1.2 Quality and Reliability Awareness

Ever since the invention of the transistor in 1940s, the semiconductor industry has grown into diverse applications areas. These range from entertainment electronics on one side to space applications on the other side. Computers and tele-communication are other notable applications. Irrespective of the application areas, the quality and reliability demands for semiconductor devices have significantly increased [13,14]. This requirement is not difficult to understand.

It is a well known rule of thumb that if it costs *one* dollars to test a defective component at chip level, it will cost *ten* dollars at board level and *hundred* dollars at system level to test, diagnose and replace the same defective component. Therefore, economically it makes a lot of sense to build a system with high quality components. Mustafa Pulat and Streb [21] put numbers into this hypothesis to make it clear: *Imagine a process step that provides 99% product yield generates 10,000 defective parts per million items produced. If 30 process steps are required to make a product, each with 99% yield, assuming independence, the overall product yield would be only 74%.* Hence, modest failure rates at the component level may result in significant likelihood of failure at the board or system level. The increasing system complexities require still better quality from IC suppliers so as to make economic products.

On the other hand, market economics forced what were known as purely digital integrated circuits to incorporate embedded memories as well as analog blocks so as to offer cheaper and reliable system solutions. Today, very often a single IC package would include all functional blocks of an entire micro-controller or digital signal processor (DSP) including the memories and analog interfaces. These functional blocks or system components now must share a common substrate and manufacturing process. This resulted in a dramatic change in the case of memories - especially DRAM, and analog circuits, which very often must be fabricated on a process developed just for standard logic. High packing density, standard manufacturing process implementation and dynamic nature of operation make such circuits (DRAMs, analog blocks) susceptible to a variety of manufacturing process defects. These developments, though far reaching in terms of market penetration, caused anxiety amongst design, process and test professionals.

As systems became more complex, their upkeep, maintenance and repair became costlier. Often specialists are required for such functions. Therefore, reliable system operation over its life-time became another absolute requirement. These developments led to slogans like *Design for Quality* and *Design for Reliability.* The terms quality and reliability are often misunderstood. Here, for the sake of clarity, we must distinguish between the terms quality and reliability. According to Hnatek [14]: *The words "reliability" and "quality" are often used interchangeably as though they were identical facets of a product's merit; however, they are different. Quality pertains to the population of faulty devices among the good ones as they arrive at the user's plant. Or, in another view, quality is related to the population of faulty devices that escape detection at the supplier's plant.... Reliability is the probability that an IC will perform in accordance with expectations for a predetermined period of time in a given environment. Thus reliability is quality on a time scale, so to speak, and testing (screening) compresses the time scale.*

1.3 Building Quality and Reliability

Design, fabrication process and test form three major activities in the development of an IC. It is futile to believe that overall quality of any IC can be achieved considering only design, process or test alone. In other words, robust design, controlled process and effective test together result in a quality product.

Role of design and process in building IC quality and reliability has been investigated in depth and is the focus of further investigations [31]. From the manufacturing standpoint, process and device technologies in the deep-submicron region (0.25-0.10 μ m) are, however, approaching practical limits and therefore concurrent achievements in high performance, high packing density, and high reliability are expected to become increasingly difficult. Besides, quality and reliability issues for VLSI (with as many as 10^9 transistors on a chip) are becoming more stringent due to required escape rate of less than 10 parts per million (PPM) and required failure rate of less than 10 FIT [6,25]. One device failure in 10^9 device operating hours is termed as one FIT. Furthermore, due to large initial investment required by the fabrication process complexity, it has recently become a matter of considerable debate whether such an investment is profitable. Similarly, contribution of design to improve quality and reliability of ICs has been outstanding and beyond the objectives of this book.

1.3.1 Role of Testing in Quality Improvement

The often stated objective of testing is to ensure the quality of the designed systems. Testing is the last check-post before the product is shipped to its destination. In other words, it is the last opportunity to prevent the faulty product from being shipped. Mustafa Pulat and Streb [21] stressed the need of component (IC) testing in total quality management (TQM). In a large study spreading over three years and encompassing 71 million commercial grade ICs, Hnatek [14] reported differences in quality seen by IC suppliers and users. One of the foremost conclusions of the study was that IC suppliers often do not do enough testing. How thorough must functional testing of digital ICs be to guarantee adequate quality? Is fault grading necessary? If yes, how high must the single-stuck fault coverage be for a given quality? These were the objectives of a study conducted by Agrawal et al. [2]. They described a model based technique for evaluating the fault coverage requirement for a given field escape rate (PPM). In their subsequent paper [27], authors showed that the fault simulation results with tester data can also predict the yield and fault coverage requirements for a given PPM for an IC. It was shown that for 1000 PPM about 99% fault coverage will be needed. Similar results were obtained by McCluskey and Buelow [19]. The result of their theoretical analysis as well as one experimental evidence indicated that logic production test fault coverage of greater than 99% is necessary for manufacturing and selling high quality ICs.

At the same time, it was discovered that classical voltage based test methods for digital CMOS ICs are grossly inadequate in ensuring the desired quality and reliability levels [8,22]. Many commonly occurring defects like gate oxide defects often are not detected by logic tests [12,30]. Therefore, such escaped defects are quality and reliability hazards. This increased quality awareness brought in new test techniques like quiescent current measurements (QCM), or I_{DDQ} as it is popularly known, in the test flow for digital CMOS ICs [4,11,19,20,22,29]. Arguably I_{DDQ} is the most effective test method in catching manufacturing process defects. Perry [22] reported that with the implementation of I_{DDQ} testing on ICs, the system failure rate dropped by a factor

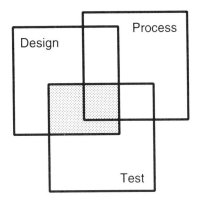

Fig. 1.1: Non-idealized IC realization process.

of six. Gayle [10] reported that with implementation of I_{DDQ} testing the defect rate had fallen down from a high 23,000 parts per million to a more acceptable 200 parts per million. Similarly, Wiscombe [33] reported improvement in quality levels.

Analog test complexities are different from that of digital circuits. The application of digital design for testability (DfT) schemes have been largely unsuccessful in analog domain [24]. As a result, a vast majority of analog circuits are tested by verifying the functionality (specifications) of the device. Since different specifications are tested in different manners, it makes analog functional testing costly and time consuming. Moreover, often extra hardware is needed to test various specifications. Limited functional verification does not ensure that the circuit is defect-free and escaped defects pose quality and reliability problems. Defect oriented testing is expected to provide a structured analog test methodology which will improve the quality, reliability and economics of tested devices.

1.4 Objectives of this book

Design, fabrication process and test constitute three major steps in the realization of an IC. In an idealized environment these three steps should perfectly match. For example, an ideal design realized in an ideal fabrication process environment will have 100% yield. Therefore, test in an ideal environment is redundant and not required. The real world is far from an idealized one where all these steps have a certain amount of uncertainty associated with them. Fig. 1.1 symbolically illustrates the non-idealized IC realization process with the three major steps having partial overlap with each other. The partial overlap signifies an imperfect relationship amongst the steps. In other words, only a subset of fabricated ICs are defect-free and only a subset of defective ICs are caught by the test. As a result, a set of design, test, and process professionals have to make a conscious effort to strive for near optimum relationship for better product quality and economics. For example, the test should cover all the likely defects in the design, or the design should work within the constraints of the process, or the test should incorporate the process defect information for optimum utilization of resources.

In this broad spectrum, this book focuses on the darkened area of Fig. 1.1. The primary objective of the book is to make readers aware of process defects and their impact on test and quality. The target audience of this book is VLSI design and test professionals. The motivation of the book comes from the fact that costs of IC testing have risen to absurd levels and are expected to rise further for systems on chip. According to experts, design and test professionals have to focus on defects rather than higher level fault models to reduce the test cost while improving the quality and reliability of products. It is a daunting task given the complexity of modern ICs. Furthermore, shrinking technology makes circuits increasingly prone to defects. Millions of dollars are spent in state of the art manufacturing facilities to reduce particulate defect count (primary cause of yield loss), defect monitoring, yield improvement, etc. Therefore, in such a scenario, knowledge of what can go wrong in a process should be advantageous to design and test professionals. This awareness can lead to robust design practices such that the probabilities of many types of defects are reduced or, alternatively, their detection is simplified. Similarly, test solutions for dominant types of defects may be generated to rationalize test costs.

There are a number of defects types that may occur in a circuit and often different circuit types have to co-exist on the same die. Depending upon the circuit type (dynamic, static, digital, RAM, PLAs or analog) defects influence the operation differently. Hence, such circuits should be addressed separately and optimum test solutions for each circuit type should be evolved. For example, certain classes of defects are not detected by logic testing, however, are readily detected by I_{DDQ}. A good DfT scheme is the one that works within the constraints of a given circuit type. A few of these schemes are suggested in subsequent chapters and may be used to create test modes such that the defect coverage of the test is enhanced or very few tests are needed for defect detection.

Why this book? The field of defect oriented testing is nearly two decades old. The information on defect oriented testing is fragmented and distributed in various conference and journal papers. There is hardly any book providing a cohesive treatment this field deserves. In this book an attempt is made to bridge this gap and provide an overview of this field. This book differs in concept and organization from recently published book on contamination, defects, and faults [16]. Our focus in this book is to study the impact of defects on various circuit types and draw conclusions on defect detection strategies. This book does not pretend to include all the work done in this area. However, an effort is made to include most of practically relevant information in this area and present it in a readable format. The book is written keeping practical VLSI aspects in mind. The DfT strategies described in the book are realizable in CMOS technology and many have actually been implemented at Philips Semiconductors and elsewhere.

1.5 Outline of the book

In the following chapter, a review of digital fault models is provided. These fault models are classified according to the level of abstraction. The relative merits and shortcomings of these methods are reviewed. The differences between functional and structural testing are brought out and impracticality of functional testing for complex VLSIs is highlighted. The second half of the chapter is devoted to the fundamental concepts of defects, faults and the inductive fault analysis (IFA) technique for realistic fault model development. Relationship between process deformations and IC failures is illustrated.

In Chapter 3, some of the earlier work on defects in simple NMOS and CMOS logic circuits are studied. Some of the earlier studies on the effectiveness of the stuck-at (SA) fault model in detecting defects in CMOS circuits are discussed and their conclusions are summarized. The early work on gate oxide defects and its impact on IC quality and reliability is presented. Often such defects are not detected by voltage testing and I_{DDQ} measurements are needed to detect them. Subsequently, the studies on Boolean and I_{DDQ} testing are described and important conclusions are highlighted. Finally, enhanced leakage current and delay effects of realistic defects in CMOS circuits are illustrated.

Chapter 4 is devoted to defects in sequential circuits containing flip-flops and scan chains. Flip-flops are very important building block in a VLSI and often are tested with the help of one (or multiple) scan chains. We outline a DfT technique describing how defects in the complete scan chain may be tested with a relatively small number (4 or 6) of test vectors. The number of test vector are independent of the number of flip-flops in the scan chain. This is achieved by making the scan chain transparent in the test mode. The scan path transparency reduces the voltage test complexity of scan paths significantly and substantially improves the fault coverage of the I_{DDQ} test. A set of better testable flip-flop configurations are illustrated. These configurations also have improved performances compared to standard flip-flop configurations.

Random access memories (RAMs) are integral parts of modern ICs as well as systems. Proliferation of microprocessor, DSP, micro-controller based systems require large amount of embedded and dedicated RAMs. As far as their testing is concerned, RAMs suffer from quantitative issues of digital testing as well as qualitative issues of analog testing. In Chapter 5, we address the application of defect oriented test method to RAMs. The application of this method results in efficient algorithms whose effectiveness is demonstrated with silicon test data. RAMs are also endowed with a fairly regular and well defined architecture. Therefore, an intelligent mix of I_{DDQ} and voltage testing has potential of reducing RAM test complexity significantly compared to any conventional (voltage based) RAM test algorithm. A DfT strategy is described such that all bridging defects in the RAM matrix, including the gate oxide defects, are detected by four I_{DDQ} measurements. For other defects, the I_{DDQ} test is supplemented with a voltage based march test to detect the defects not detectable by I_{DDQ} testing [23]. The overall test complexity of the combined test is reduced for a given realistic fault model.

Chapter 6 is devoted to testing of defects in programmable logic devices (PLD). Such circuits are widely used in VLSI circuits and systems since they provide a simple and quick method of implementing complex Boolean functions. Recent resurgence of programmable logic devices (PLDs) ushers a new era in digital system design [5]. Modern PLD architectures offer flexibility of logic implementation. However, the same flexibility results in enormous test complexity of these devices. In this chapter, a brief review of testable PLA architectures is presented and a methodology for testing defects is outlined. Subsequently, I_{DDQ} testable PLAs are described.

In Chapter 7, the defect oriented test methodology is applied to find out non-specification based analog test methods. Owing to the non-binary nature of their operation, analog circuits are influenced by process defects in a different manner than digital circuits. In this chapter, we demonstrate with the help of a real CMOS circuits that simple test stimuli, like DC, transient and AC, can detect most of the modeled process defects. This test methodology is structured and simpler and, therefore, results in substantial test cost reduction. Furthermore, we tackle the issue of analog fault grading. The quality of the test, and hence the tested device, depends heavily on the defect (fault) coverage of the test vectors. Therefore, it is of vital importance to quantify the fault coverage. We demonstrate how the IFA technique can be exploited to fault grade given (conventional) test vectors. Once, the relative fault coverage of different blocks is known for given test vectors, an appropriate DfT scheme can be applied to the areas where fault coverage of existing test methods is relatively poor.

Finally, in Chapter 8 conclusions on defect oriented testing are given. Its advantages and limitations are outlined. Some potential research directions are recommended.

References

1. M. Abadir and A.P. Ambler, *Economics of Electronic Design, Manufacture and Test,* Boston: Kluwer Academic Publishers, 1994.

2. V.D. Agrawal, S.C. Seth and P. Agrawal, "Fault Coverage Requirement in Production Testing of LSI Circuits," *IEEE Journal of Solid State Circuits*, vol. SC-17, no.1, pp. 57-61, February 1982.

3. A.P. Ambler, M. Abadir and S. Sastry, *Economics of Design and Test for Electronic Circuits and Systems,* New York: Ellis Horwood, 1992.

4. K. Baker, "QTAG: A Standard for Test Fixture Based I_{DDQ}/I_{SSQ} Monitors," *Proceedings of International Test Conference*, 1994, pp. 194-202.

5. S.D. Brown, *"Field-Programmable Devices: Technology, Applications, Tools,"* 2nd Edition, Los Gatos: Stan Baker Associates, 1995.

6. D.L. Crook, "Evolution of VLSI Reliability Engineering," *Proceedings of International Reliability Physics Symposium*, 1990, pp. 2-11.

7. C. Dislis, J.H. Dick, I.D. Dear and A.P. Ambler, *Test Economics and Design For Testability,* New York: Ellis Horward, 1995.

8. F.J. Ferguson and J.P. Shen, "Extraction and Simulation of Realistic CMOS Faults using Inductive Fault Analysis," *Proceedings of International Test Conference*, 1988, pp. 475-484.

9. A.V. Ferris-Prabhu, "Computation of the critical area in semiconductor yield theory," *Proceedings of the European Conference on Electronic Design Automation, 1984*, pp.171-173.

10. R. Gayle, "The Cost of Quality: Reducing ASIC Defects with I_{DDQ}, At-Speed Testing and Increased Fault Coverage," *Proceedings of International Test Conference*, 1993, pp. 285-292.

11. R. Gulati and C. Hawkins, I_{DDQ} *Testing of VLSI Circuits,* Boston: Kluwer Academic Publishers, 1993.

12. C. Hawkins, and J. Soden, "Reliability and Electrical Properties of Gate Oxide Shorts in CMOS ICs," *Proceedings of International Test Conference*, 1986, pp. 443-451.

13. E. R. Hnatek, "IC Quality - Where Are We?" *Proceedings of International Test Conference*, 1987, pp. 430-445.

14. E. R. Hnatek, *Integrated Circuits Quality and Reliability,* New York: Marcel Dekker, Inc., 1987.

15. M. Inoue, T. Yamada and A. Fujiwara, "A New Testing Acceleration Chip for Low-Cost Memory Test," *IEEE Design and Test of computers*, vol. 10, pp. 15-19, March 1993.

16. J. Khare and W. Maly, *From Contamination to Defects, Faults and Yield Loss,* Boston: Kluwer Academic Publishers, 1996.

17. P. Mazumder and K. Chakraborty, *Testing and Testable Design of High-Density Random-Access Memories,* Boston: Kluwer Academic Publishers, 1996.

18. W. Maly and M. Patyra, "Design of ICs Applying Built-in Current Testing," *Journal of Electronic Testing: Theory and applications*, vol. 3, pp. 397-406, November 1992.

19. E.J. McCluskey and F. Buelow, "IC Quality and Test Transparency," *Proceedings of International Test Conference*, 1988, pp. 295-301.

20. S. D. McEuen, "IDDq Benefits," *Proceedings of VLSI Test Symposium*, 1991, pp. 285-290.

21. B. Mustafa Pulat and L. M. Streb, "Position of Component Testing in Total Quality Management (TQM)," *Proceedings of International Test Conference*, 1992, 362-366.

22. R. Perry, "I_{DDQ} testing in CMOS digital ASICs," *Journal of Electronic Testing: Theory and applications*, vol. 3, pp. 317-325. November 1992.

23. M. Sachdev, "Reducing the CMOS RAM Test Complexity with I_{DDQ} and Voltage Testing," *Journal of Electronic Testing: Theory and Applications,* vol. 6, no. 2, pp. 191-202, April 1995.

24. M. Sachdev, "A Defect Oriented Testability Methodology for Analog Circuits," *Journal of Electronic Testing: Theory and Applications*, vol. 6, no. 3, pp. 265-276, June 1995.

25. A. Schafft, D. A. Baglee and P. E. Kennedy, "Building-in Reliability: Making it Work," *Proceedings of the International Reliability Physics Symposium*, 1991, pp. 1-7.

26. Semiconductor International, vol. 14, no. 5, pp. 62, May 1991.

27. S.C. Seth and V.D. Agrawal, "Characterizing the LSI Yield Equation from Wafer Test Data," *IEEE Transactions on Computer-Aided Design*, vol. CAD-3, no. 2, pp. 123-126, April 1984.

28. P. Singer, "1995: Looking Down the Road to Quarter-Micron Production," Semiconductor International, vol. 18, no. 1, pp. 46-52, January 1995.

29. J.M. Soden, C.F. Hawkins, R.K. Gulati and W. Mao, "I_{DDQ} Testing: A Review," *Journal of Electronic Testing: Theory and applications,* vol. 3, pp. 291-303, November 1992.

30. M. Syrzycki, "Modeling of Spot Defects in MOS Transistors," *Proceedings of International Test Conference,* 1987, pp. 148-157.

31. E. Takeda et al., "VLSI Reliability Challenges: From Device Physics to Wafer Scale Systems," *Proceedings of IEEE*, vol. 81, no.5, pp. 653-674, May 1993.

32. P. Varma, A.P. Ambler and K. Baker, "An Analysis of The Economics of Self-Test," *Proceedings of International Test Conference*, 1984, pp. 20-30.

33. P. Wiscombe, "A Comparison of Stuck-At Fault Coverage and I_{DDQ} Testing on Defect Levels," *Proceedings of International Test Conference*, 1993, pp. 293-299.

CHAPTER 2

Digital CMOS Fault Modeling and Inductive Fault Analysis

We begin with an overview of digital fault models. Different fault models are classified according to the level of abstraction. The merits and shortcomings of these models are reviewed. The second half of the chapter is devoted to the defect oriented fault modeling methodology or Inductive Fault Analysis, as it is popularly known. Unlike the conventional fault modeling methods, IFA takes into account the circuit layout and manufacturing process defects to generate realistic and layout dependent faults.

2.1 Objectives of Fault Modeling

The exponential increase in the cost of functional testing has led to tests that are not functional in nature, but are aimed to detect the possible faulty conditions in ICs. The circuit under test (CUT) is analyzed for faulty conditions and tests are generated to detect the presence of such conditions. Like any other analysis, this fault analysis also requires a model (or abstraction) to represent the likely faults in ICs with an acceptable level of accuracy. This type of model is called the fault model and this type of testing is known as the structural testing. The name structural test comes from two counts. First, the testing is carried out to validate the structural composition of the design rather than its function and, second, the test methodology has a structured basis, i.e., the fault model for test generation. In fact, the concept of structural testing dates back to the 50s. In one of the first papers on the subject, Eldred proposed a methodology which will test whether or not all tubes and diodes within a gating structure are operating correctly [19]. However, structural testing gained popularity in the 70s and the 80s when structural design for test (DfT) methodologies like scan path and level sensitive scan design (LSSD) [18,24] emerged. These DfT methods became popular because their application could change distributed sequential logic elements into a big unified shift-register for testing purposes [18,24]. As a result, the overall test complexity is reduced [96]. Owing to these techniques, the test generation and fault grading for complex digital circuits became a possibility.

Breuer and Friedman [9] described fault modeling as an activity concerned with the systematic and precise representation of physical faults in a form suitable for simulation and test generation. Such a representation usually involves the definition of abstract or logical faults that produce approximately the same erroneous behavior as the actual physical defects. Here, it is important to distinguish between a defect and a fault. A defect is physical in nature and a fault is its representation. Therefore, a fault can also be defined as follows: A *fault is the electrical impact of a physical defect at an appropriate level of abstraction.* A Fault is often represented by its simulation model which is termed as the fault model. Fault models have played a pivotal role in the success of the structural testing whose goal is to test for the modeled faults. Structural testing has some notable advantages over functional testing. Foremost amongst them are:

- The effectiveness of the structural test is quantifiable. It is possible to ascertain the percentage of the modeled faults tested by a given test suite. This percentage is popularly known as the fault coverage. Thus, it allows the user to establish a relationship between the fault coverage of the test suite and the quality of tested ICs.

- Test generation for structural tests is considerably simpler compared to functional test generation for a complex CUT. Computer aided design

(CAD) tools (e.g., automatic test pattern generator (ATPG) and fault simulator) ensure faster and effective test generation.

- In the case of functional testing the choice of ATE is closely related to the CUT specifications that may cause an undesirable dependence on a particular ATE for testing. However for structural testing, the choice of ATE is largely independent of the CUT specifications which allows a greater freedom in choosing an ATE.

The underlying assumption behind structural testing is that the design is essentially correct and its function on silicon has already been verified and characterized. The non-ideal manufacturing process introduces defects in the design (or IC). These defects cause faults which result in erroneous IC behavior that needs to be tested. Moreover, fault models in structural testing are only for the fault simulation purpose and they do not represent how likely a fault is compared to another one. Structural testing further assumes the time invariance of a fault. Time variant faults degrade over time, and therefore, are difficult to test in the production environment. Similarly, a combination of environmental conditions may trigger a device to have a temporary fault that is very difficult to test in the production environment. Therefore, time variant and temporary faults are not considered for structured test generation. Lastly, it is assumed that a fault has a local impact. For example, a fault may cause the output of a NAND gate to be always logic high. The origin of this assumption lies in the fact that a fabrication process line is regularly monitored, hence, the global defects are controlled early in the production environment. As a result, the vast majority of defects which are to be tested for are local in nature.

2.2 Levels of Testing

It is prudent to mention here that the aims of the structural test are different from those of the functional test. Beenker et al. [6] suggested that test objectives should be mapped to various product development stages, and hence have different purposes. Different levels of testing and their significance are explained with the help of Fig. 2.1. The product development stages range

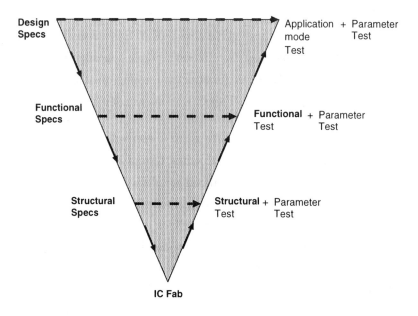

Fig. 2.1: An idealized verification process [6].

from requirement specification to functional specification, functional design, and structural design [6]. These stages are highlighted on the left hand side of the figure. The corresponding test levels, called *verification tests*, are shown on the right hand side of the figure. Each has a different objective, namely, assuring the corresponding specifications are faithfully implemented. At the end of the verification process, if the design has no flaws, it is passed on to large scale production and the verification tests are replaced by the production tests.

As mentioned before, objectives of structural testing include testing for the faithfulness of the structure of the IC. Fault models are required which represent production process defects. These faults are mapped onto the device under test (DUT) and test vectors are generated to cover such failures. Often it is not possible, due to economic considerations, to test for all possible defects in the DUT. Usually a compromise is reached between the economic considerations and the type of fault model. DfT schemes (e.g., scan path, LSSD and macro test) provide test solutions while retaining/improving fault coverage economically.

Functional testing contrasts with the above mentioned approach. Its objective is to test the function of the DUT. The functional test vectors are not based on a fault model but focus on the function of the DUT. Often, functional test vectors are provided by the designer. The quality of a functional test can not be objectively quantified.

However, the correct design of a device alone does not ensure that the device will work satisfactorily in the system. Increasing complexities of ICs and systems forced test professionals to find methods of characterizing the behavior of ICs in their real application. The application mode testing emerged from this desire. During this type of testing the ICs are tested with real life data in a wide variety of environmental conditions that the DUT is supposed to encounter during its normal life cycle. In this fast moving world being good is not good enough. How good is the good!! is one of the questions answered by parametric testing. In other words, quantifying the performance of the DUT in terms of speed, AC and DC parameters, power consumption, and environmental (i.e., temperature and voltage) susceptibility are the objectives of parametric testing.

2.3 Levels of Fault Modeling

Numerous possible ways exist to represent faults for the purpose of fault simulation. In general, faults are categorized according to the level of abstraction. The correct level of abstraction essentially achieves a trade-off between the fault model's ability to accurately represent an actual physical defect and the speed of fault simulation. For example, behavior (function) level fault modeling is the fastest but is least accurate. On the other hand, layout level inductive fault analysis (IFA) is most accurate but requires enormous computational resources. Most fault models can be classified according to the following levels of abstraction.

2.3.1 Logic Level Fault Modeling

Initial work on fault modeling was concentrated at the logic level. It was assumed that the faulty behavior due to defects can be mapped onto the incorrect Boolean function of basic gates in the CUT. Simple circuits and relatively large feature sizes justified this assumption. Moreover, fewer defects could cause fatal faults. In the early days, the semiconductor industry was battling to solve complex design and process related problems and paid little attention to IC testing. Yield of the early ICs was poor. The poor yield was caused primarily by equipment and technological problems and not by the spot or lithographic defects. Furthermore, limited knowledge about the origin of defects and the impact of defects on circuit behavior forced researchers to adopt many simplifying assumptions in test.

The implementation details of logic gates are not considered in fault modeling at logic level. Fault modeling at the logic level has some notable advantages. The Boolean nature of the fault model allows the usage of powerful Boolean algebra for deriving tests for complex digital circuits [2]. The gate level representation of the faulty behavior resulted in a technology independent fault model and test generation algorithms. Technology independent tests increased the ability to port designs to different technologies.

2.3.1.1 Stuck-At Fault Model

The stuck-at fault (SAF) model is the most commonly used logic-level fault model. Poage [59] was one of the first to propose the SAF model. It became a popular fault model in the 1960s owing to its simplicity. It is widely used in academic research as well as in the industry. Its simplicity is derived from its logical behavior. SAFs are mapped onto the interconnects (or nets) between logic gates. Thus, they are also referred to as stuck-line (SL) faults [31,32]. Under the faulty condition, the affected line is assumed to have a permanent (stuck-at) logic 0 or 1 value that can not be altered by input stimuli. Fig. 2.2 illustrates a NAND gate and its truth table. Let us consider that the line A has a stuck-at-1 (SA1) fault. The presence of the SA1 fault is detected by the faulty

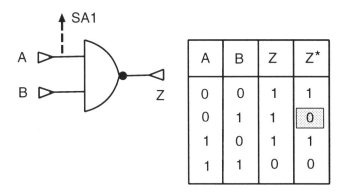

Fig. 2.2: The SAF model and its fault-free and faulty Boolean behavior.

gate response when lines A and B are driven to logic 0 and 1, respectively. A fault is said to be detected when the expected output of the logic gate differs from the actual output. For example, the third and fourth columns of the table in Fig. 2.2 illustrate the expected (fault-free, Z) and the actual (faulty, Z*) responses of the NAND gate. For the test vector A=0, B=1, the expected and actual responses differ and thus the fault is said to be detected.

The SAF model is widely used till date for many practical reasons. These reasons include the availability of computer aided design (CAD) tools for test generation, and the technology independence of the SAF model. Furthermore, many physical defects cause SAFs. With increasingly complex devices and smaller feature sizes, the likelihood that more than one SAF can occur at the same time in an IC has become significantly large. However, a large number of possible multiple SAFs force the test generation effort to be impractical. For example, assume that a circuit contains n lines. Each line is capable of having three distinct states; i.e, SA0, SA1 and fault-free. Therefore, there are $3^n - 1$ multiple SAFs possible. A typical IC may contain hundred thousands or more lines which may result in an enormously large number of possible faults. Therefore, it is a common practice to assume the occurrence of only single-stuck-line (SSL) faults in the IC. Hence, for an IC with n lines, only $2n$ SSL faults are possible.

Table 2.1: The SAF classes in a 2 input NAND gate.

A	B	Z	Fault Classes
1	1	0	A/0, B/0, Z/1
1	0	1	B/1, Z/0
0	1	1	A/1, Z/0
0	0	1	Redundant Test

2.3.1.2 Fault Equivalence, Dominance and Collapsing

A fault f1 is considered to be equivalent to fault f2, if their faulty behaviors for all possible input stimuli are indistinguishable from each other at primary outputs. Therefore, a stimulus detecting f1 will also detect f2. Such faults can be bunched into an equivalent fault class. This point can be further illustrated from Fig. 2.2. Consider once again a 2-input NAND gate. A SA0 fault at B (B/0) causes a SA1 fault at Z (Z/1) for all input conditions of the NAND gate. Therefore, both of these faults are indistinguishable from each other at the primary outputs. A prior knowledge of the fault classes in a network is useful in fault diagnosis. Furthermore, fault detection is simplified by using equivalent fault classes to reduce or collapse the set of faults into fault classes that need to be considered for test generation [51,75]. The fault resolution of a network depends on how widely equivalent faults are separated. Hence, knowledge of the equivalent fault classes is helpful in problems such as test point placement and logic partitioning to increase the fault resolution [30].

Poage [59] was the first to describe the concept of fault dominance. According to him, if all tests for a fault f1 also detect f2 but only a subset of tests for f2 detect f1, then f2 is said to dominate f1. However, Abraham [2] argued that it is safer to consider a test for f1 which will ensure both the faults are detected.

Hence, f1 is said to dominate f2. For example, in Table 2.1, A/1 does not cause Z/0 behavior for all input conditions of the NAND gate. A test for A/1 will also test for Z/0. Converse is not true because Z/0 is also tested by the test for B/1. Hence according to Abraham, the detection of A/1 will ensure the detection of Z/0 as well. Therefore, A/1 is said to dominate Z/0. On the other hand, according to Poage, Z/0 dominates A/1. However, it should be mentioned here that the difference between two approaches is in the definition and not in the process of fault collapsing. The concepts of fault equivalence and fault dominance allow us to collapse SAFs into fault classes. In general, for an n-input gate there will be $n+1$ equivalent fault classes [2]. Table 2.1 shows these fault classes and test vectors needed for their detection for a 2-input NAND gate.

These concept can be applied to larger circuits [23,30,35,51,68,75,77,89], but in general it is a complex computational problem. It is shown [30,35] that the problem of identifying fault equivalence in arbitrary networks belongs to the class of computationally difficult problems called NP complete. Nevertheless, a significant attention has been devoted to find equivalent fault classes in circuits bigger than a single logic gate [23,30,51,68,75,77,89].

Schertz and Metze [75] described fault collapsing as the process of combining faults by means of implication relationships derived from the network. They defined three stages of fault collapsing corresponding to three types of implication relationships. These stages can be explained as follows: Consider a fault table $[T_{ij}]$ that contains a row for each possible input vector and each fault is represented by a column. The entry t_{ij} is 1 if and only if test i detects fault j. A column j dominates column k if for every 1 in column k there is also 1 in column j (same as definition by poage). If two columns of the table are identical then each dominates the other. The first stage of the fault collapsing corresponds to identical columns in the fault table. The second stages of collapsing corresponds to the unequal columns, that is, the situations where one fault is more readily detected than the other. The third stage is concerned with the relationship between single and multiple faults. In order to illustrate these stages, let us consider a two input NAND gate of Fig. 2.2 and its fault classes in Table 2.1 once again. The faults A/0, B/0, and Z/1 are indistinguishable from each other and, hence, represent the first stage collapsing procedure (first fault class in Table 2.1). The next two rows of the table represent the second stage of fault collapsing. The third stage of fault collapsing is concerned with the multiple faults. For example, Z/0 is indistinguishable from all input lines (A/1, B/1) having a multiple SA1 fault.

In the case of larger circuits having non-reconverging fanouts, the application of foregoing analysis is rather straight forward. However, in the case of reconverging fanouts, the reconverging branch may interfere with the propagation of the fault to a primary output. Gounden and Hayes [30] further simplified the fault class identification problem for certain cases. They introduced the concepts of intrinsic and extrinsic fault equivalence. These concepts were utilized to derive some general conditions for fault equivalence and non-equivalence for a given network topology. They argued that every switching function can be realized by a two-level network. Therefore, for a CUT realized by a two-level network, it is possible to identify equivalent fault classes. For a given two-level network, the computational complexity is reduced to 19% of the original value [30].

Fault collapsing techniques have been extended to NMOS and CMOS circuits not only for stuck-at faults but also for transistor stuck-open (SOP) and stuck-on (SON) faults [23,77].

2.2.1.3 Mapping of CMOS Defects on Stuck-At Faults

In CMOS technology an m-input logic gate is realized with m p-channel transistors and an equal number of n-channel transistors. The output of the gate is taken where p and n-channel transistor groups are connected to each other. Depending upon the Boolean state of inputs, the output is driven either by p-channel transistor(s) to logic 1 or by n-channel transistor(s) to logic 0. The output is never driven by both type of transistors at the same time.

Fig. 2.3 illustrates a transistor level schematic of a 2-input NAND gate in the CMOS technology. Two p-channel and two n-channel transistors are needed

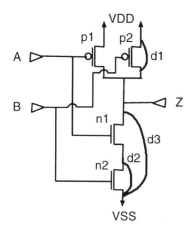

A	B	Z	Defect Detection
1	1	0	d1
1	0	1	d2, d3
0	1	1	d3
0	0	1	Redundant Test

Fig. 2.3: Defects in a 2 input CMOS NAND gate and their detection.

to realize a 2-input NAND gate. The figure also illustrates some common bridging defects in the gate and the table shows how they are detected by SAF test vectors. In this simplistic analysis we assume that the defect resistance is substantially low compared to the "on" resistance of a transistor. Therefore, in case of a conflict between defect and a transistor, the defect will override. Later, we will remove this assumption as we discuss lower level fault models.

Defect d1 causes a short between the output Z and power supply VDD. Assuming that the on-resistance of n1 and n2 transistors is substantially high compared to defect resistance, the defect results in the Z/1 stuck-at fault. Needless to say that A=1, B=1, test vector will detect it. Similarly, defect d3 causes the Z/0 stuck-at fault which can be detected by A=0, B=X, or A=X, B=0, test vectors.

Defect d2 can not be modeled by the SAF but it can be detected by SAF test vectors under the assumption that resistance of the defect and the on-resistance of the n1 transistor is substantially smaller than the on-resistance of the p2 transistor. Test vector A=1, B=0 causes transistors p2 and n1 to conduct. Therefore, in the presence of d3 output Z is driven from VDD and VSS simultaneously and has an intermediate voltage. This intermediate voltage is not the same as Z/1 or Z/0. However, this defect is detected provided that the following logic gate interprets this output level as a logic 0 instead of a logic 1.

2.3.1.4 Shortcomings of the Stuck-At Fault Model

In spite of its simplicity and universal applicability, the SAF model has some serious drawbacks in representing defects in CMOS technology. It can represent only a subset of all defects. Large number of defects that are not detected by a SA test set cause bad ICs to pass the test. In a study, Woodhall et al. [97] reported that open defects led to an escape rate of 1210 PPM when the CUT was tested with a 100% SAF test set. We illustrate some representative examples of faults that are not modeled by SAF.

(i) Open Defects

The output of a CMOS logic gate retains its value when left in high impedance state. Such a property has found numerous applications in data storage and time discrete signal processing area. Furthermore, CMOS circuits offer very high input impedance, hence, floating interconnects retain their previous logic value for a significantly long time. This sequential behavior of CMOS logic gates causes many open defects not to be detected by a SAF test set [95].

Consider once again the same 2-input NAND gate and its truth table illustrated in Fig. 2.3. Some of the open defects affecting the operation of transistor p1 are not detected. The second test vector (A=1, B=0) drives the output to logic 1 through transistor p2. The third test vector (A=0, B=1) instead drives the output to logic 1 through transistor p1. Some of the open defects affecting transistor p1 can not be detected by the third test vector (Fig. 2.4). In the presence of these defects, the output is not driven to logic high in the third test vector but retains its logic high state from the second test vector. Therefore, these defects are not guaranteed to be detected by SA test vector set. However, these defects are detected if the order of test vectors is changed. For example, test vectors (T2, T1, T3) will detect these faults. A detailed treatment of open defects is presented in the next subsection. However, in general, for open defects two test vectors (T1,T2) are required. The first test vector T1 initializes the output to the a logic level and T2 attempts to change the output state through a

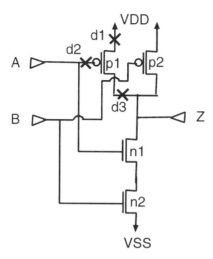

Fig. 2.4: Undetected open defects in a 2 input CMOS NAND gate by the SAF test vectors.

particular transistor-controlled path. For logic gates with higher complexity the SAF test vector set can not guarantee detection of all open defects.

(ii) Short Defects

A short defect is defined as an unintended connection between two or more otherwise unconnected nodes. Often they are referred to as bridging faults or simply as bridges. Shorts are the dominant cause of faults in modern CMOS processes. In the CMOS technology, shorts can not be modeled as wired-OR or wired-AND logic. The circuit level issues (e.g., W/L of driving transistors, defect resistance, logic thresholds of subsequent logic gates, Boolean input logic conditions, etc.) play a very important role in their detection. Although a large number of shorts (shorts between VDD/VSS and input/output of logic gates) lead to SAFs, in general, the SAF model does not adequately represent all possible shorts in an IC. Hence, SAF test vectors do not ensure the detection of all shorts.

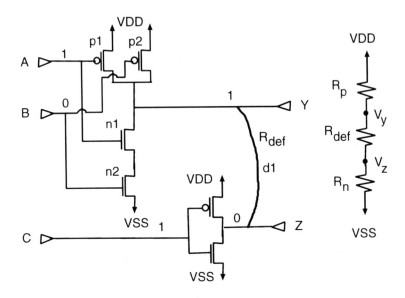

Fig. 2.5: An external bridging defect not detected by the SAF test vectors.

Shorts in ICs can be classified as internal bridges and external bridges. Internal bridges are those that affect the nodes within a logic gate. Shorts shown in Fig. 2.3 are examples of this category. The external bridges are those that affect nodes within two or more logic gates. Fig. 2.5 illustrates an external bridge and its electrical model. Besides the circuit level issues, the detection of external bridging faults also depends on exciting nodes Y and Z to opposite logic values. For a complex circuit, it is a non-trivial task. One needs to find out all potential locations for such bridging defects. Techniques like Inductive Fault Analysis can be useful in finding such locations.

2.3.2 Transistor Level Fault Modeling

The SAF model has limitations in representing defects of CMOS circuits. In general, there are many defects in CMOS circuits that may be represented by the SAF model and are detected by the SAF test set. There are other defects that are not modeled by the SAF model but are detected by chance by the SAF test set. However, there are still potentially many defects that are not modeled by the SAF model and are not detected by the SAF test set [2]. Therefore, we need transistor level fault models which represent faulty behavior with a better accuracy. However, such fault models result in significantly larger number of faults compared to that for the SAF model. Furthermore, a lot of research effort is directed towards a better understanding of defects and their influence on circuit behavior [12,20,37,38,40,63-67,95,97]. The knowledge gained resulted in improved fault models leading to efficient and effective tests, and better quality of tested ICs.

As described earlier, a static CMOS logic gate is constructed by a set of p-channel and a set of n-channel enhancement mode transistors between VDD and VSS terminals. An enhancement mode transistor in the absence of a gate to source voltage ($V_{gs}=0$) does not conduct. It conducts only when an appropriate gate to source voltage ($V_{gs}>V_T$) is applied. For example, a p-channel transistor conducts when its gate terminal is logic 0 and an n-channel transistor conducts when its gate terminal is logic 1. Therefore, a transistor can be treated as a three terminal ideal switch having a control terminal or gate, which controls the flow of data from the source terminal to the drain terminal. In the transistor-level fault modeling, physical defects are mapped onto the functioning of these switches (transistors). For a given combination of input logic values, a CMOS logic gate may have one of the following states [65]:

1. The output node is driven to VDD via one or more paths provided by conducting p-channel transistors and no conducting path from output to VSS exists through n-channel transistors.

2. The output node is driven to VSS via one or more paths provided by conducting n-channel transistors and no conducting path from output to VDD exists through p-channel transistors.

3. The output node is not driven to VDD or VSS via conducting transistors.

4. The output node is driven by both VDD and VSS via conducting transistors.

In the first two cases, the output is logic high and low, respectively. In case 3, the output is in high impedance and its current logic state is the same as its previous logic state. In case 4, the output logic state is treated as indeterminate, since the actual voltage depends on the resistance ratio of the conducting paths of p and n-channel transistors. However, logic gates are rarely designed to have a case 4 type of situation.

2.3.2.1 Transistor Stuck-Open Fault Model

The detection of transistor stuck-open (SOP) faults has been a difficult problem and has received considerable attention in the past [12,15,17,20,37,38,40,43,63-67,95,97]. In the presence of a SOP fault the affected transistor fails to transmit a logic value from its source terminal to its drain terminal. Therefore, the transistor can be treated as a switch which never closes and remains open in spite of all possible Boolean input conditions. As apparent from the name, such faults in enhancement mode transistors are primarily caused by open defects. However, short defects can also cause a transistor to have a SOP fault. Fig. 2.6 illustrates some defects causing SOP faults in a logic gate. SOP faults are classified by their location as a fault at the source (S) or drain (D) terminal (S/D-line fault) or at the gate terminal (gate-line fault) of a transistor [40]. A S/D-line fault creates a break in the data transfer path and clearly causes a SOP fault in the transistor (defects d1 and d3, Fig. 5). A gate-line fault (defect d2, Fig. 2.6) requires explanation. As stated above, in the CMOS technology enhancement-mode p and n-channel transistors are used. Enhancement transistors have the property that in the absence of any gate voltage (V_{gs}) the transistor does not conduct. Only when the gate voltage exceeds the threshold voltage, V_T, the transistor starts to conduct. Therefore, a gate-line fault may cause a transistor to be in the non-conduction mode. A short defect between the gate terminal and the source terminal (defect d4, Fig. 2.6) of the transistor forces the same voltage on both these terminals. As a result, the p2 transistor is never in the conduction mode, at the same time the same defect causes n2 transistor to be always in the conduction mode (SON fault).

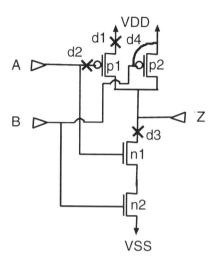

Fig. 2.6: Open and bridging defects causing the SOP fault behavior.

Wadsack demonstrated that in the presence of a SOP fault in a CMOS logic gate, the gate shows a memory effect under certain input conditions [95]. Therefore, such faults are not guaranteed to be detected by a SAF test set. In general, SOP fault detection requires a two pattern test sequence <T1, T2> [37]. The first test vector of the sequences, T1, is referred to as the initializing test vector and the second test vector of the sequence, T2, is referred to as the fault excitation test vector. Test vector T1 initializes the output to logic 0(1) and T2 attempts to set the output to logic 1(0) if the logic gate is fault-free. A failure to set the output to logic 1(0) indicates the presence of a SOP fault.

Some SOP faults in static CMOS logic gates require only one test vector, T2. If there is only one path from output to VSS (VDD) and the SOP fault affects this path, it is not possible to set the output to VSS (VDD). For example, in the case of a 2-input NAND gate (Fig. 2.4), test vector T2 (A=1, B=1) will detect SOP faults in the n-channel transistors. Effectively, such SOP faults discon-nect all possible paths from the output to VSS and cause a SA1 fault (Z/1) at

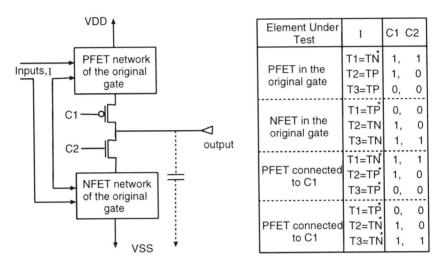

Element Under Test	I	C1	C2
PFET in the original gate	T1=TN*	1,	1
	T2=TP	1,	0
	T3=TP	0,	0
NFET in the original gate	T1=TP*	0,	0
	T2=TN	1,	0
	T3=TN	1,	1
PFET connected to C1	T1=TN*	1,	1
	T2=TP*	1,	0
	T3=TP*	0,	0
PFET connected to C1	T1=TP*	0,	0
	T2=TN*	1,	0
	T3=TN*	1,	1

Fig. 2.7: Robust SOP fault detection scheme and the 3-pattern test procedure [66].

the output of the NAND gate and fault detection requires only a single test vector [65].

(i) Design for SOP Fault Testability

A 2-pattern test for SOP fault detection can be invalidated (i.e., may not detect the fault it was supposed to detect) by arbitrary circuit delays and glitches if patterns are not carefully selected. In fact, for some irredundant CMOS complex gates a *robust* test (which is not invalidated by arbitrary circuit delays) for SOP faults does not exist [66]. At the gate level, a *robust* test sequence is a sequence of test vectors in which each successive test vector differs from previous test vector in only one bit position. However, it is difficult to generate robust test for a given logic gate from primary inputs of an IC because even a single bit position change in a test vector may produce multiple changes at the faulty gate. As a remedy for robust SOP fault detection, DfT solutions for complex gates have been proposed [66]. In the first scheme, an addition of two transistors with two independent control lines to each complex gate was suggested. Three test vectors are needed to detect a SOP fault. Fig. 2.7 shows the scheme and the test procedure. TN and TP are the fault evaluation patterns

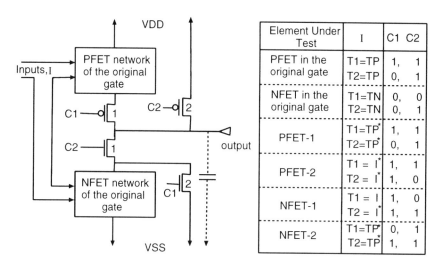

Element Under Test	I	C1	C2
PFET in the original gate	T1=TP	1,	1
	T2=TP	0,	1
NFET in the original gate	T1=TN	0,	0
	T2=TN	0,	1
PFET-1	T1=TP*	1,	1
	T2=TP*	0,	1
PFET-2	T1 = I*	1,	1
	T2 = I*	1,	0
NFET-1	T1 = I	1,	0
	T2 = I*	1,	1
NFET-2	T1=TP*	0,	1
	T2=TP	1,	1

Fig. 2.8: Robust SOP fault detection scheme and the 2-pattern test procedure [66].

(T2). The pattern TN* (TP*) is any value of input I that would have established one or more paths from VSS (VDD) to the output node.

The above mentioned scheme has a disadvantage that an SOP fault needs a 3-pattern test which may result in a longer test sequence and thus increased test costs. Therefore, Reddy et al. [66] suggested a second scheme that requires a 2-pattern test and the test will not be invalidated by arbitrary circuit delays. However, a total of 4 transistors for each complex gate are needed for implementation. The additional two transistors PFET2 and NFET2 are added between the output and VDD (VSS) in parallel to the PFET (NFET) network. The implementation of the scheme and the test procedure is shown in Fig. 2.8. To test a SOP fault in a PFET the initialization vector is provided through NFET1 and NFET2. During this time input I is the evaluation test vector, however, its evaluation is blocked by non-conducting PFET1. In the second test vector, T2, since only the control is changed and input I stays the same and the test is not invalidated by the circuit delays. Similarly, other part of a complex gate are tested for SOP faults. The test vector I* in Fig. 2.8 signifies any arbitrary input test vector. Both of these DfT schemes could also detect

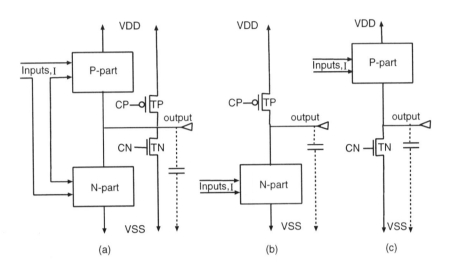

Fig. 2.9: Single pattern SOP fault detection procedure.

SOP faults in added transistors. However, practicality of these schemes is limited owing to high area overhead and performance degradation

Reddy et al. [67] presented a procedure to find the robust 2-pattern test for SOP faults in combinational circuits, if such a procedure existed, which will not be invalidated in the presence of arbitrary circuit delays. They assumed that T2 of the 2-pattern test is given and then provided a procedure to determine an appropriate initializing input T1. If no appropriate T1 is found, another test T2 is determined and then the procedure is repeated. In case the procedure fails to provide a T1 for all T2s this would imply that no robust test exists for the fault and the circuit should be redesigned to have robust SOP test.

Rajsuman et al. presented a test technique for testing of SOP faults with a single test vector [63,64]. In this technique, n-channel and p-channel transistors are tested separately. The technique is illustrated in Fig. 2.9. Part (a) of the figure shows two added transistors and two control signals for a CMOS logic gate. A full CMOS (FCMOS) logic gate is transformed into a pseudo nMOS (pMOS) gate by adding an extra high resistive pMOS (nMOS) transistor (Figs. 2.9(b) and (c)). The resistance of the transistor should be such that the output is pulled high (low) if none of the nMOS (pMOS) transistors is conducting. When single or multiple nMOS (pMOS) transistors are conducting, the output voltage is close to VSS (VDD). Two extra transistors, TP and TN, are needed and are controlled by two independent signals, CP and CN, respectively. In the normal circuit operation these transistors are switched-off (CP=1,CN=0). The testing of SOP faults in nMOS transistors is performed as follows: CP and CN are kept low. Inputs are applied to the nMOS part such that the output is pulled low through each possible Boolean combination of inputs. In the presence of a SOP fault in the nMOS part, the output is not pulled low for one or more input conditions. In fact, these inputs are the excitation vectors (T2) of the conventional 2-pattern test. Similarly, the pMOS part is tested by keeping CP and CN logic high.

In the presence of a SOP fault, the output of a pseudo nMOS (pMOS) logic gate shows a SA1 (SA0) behavior. Therefore, an automatic test pattern generating program (ATPG) can be utilized for test generation. Furthermore, a significant reduction in test generation and test application time is also expected. However, the scheme requires two transistors per logic gate. In addition, two control lines are required to control the transistors. These extra transistors themselves are untestable. Furthermore, extra transistors will cause an increase in parasitic capacitance which will have an impact on circuit performance. In a subsequent article, Jayasumana et al. [38] proposed a DfT solution which requires only one transistor and two control lines.

(ii) Layout Rules for SOP Fault Detection

In the previous sub-section, we saw that robust 2-pattern test generation for SOP fault testability is a difficult problem for a complex IC. Furthermore, SOP DfT schemes have area and performance penalties that restrict their application.

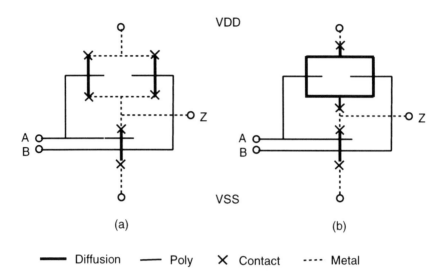

Fig. 2.10: Stick diagrams of a 2-input NAND gate; (a) conventional stick
diagram, (b) alternative stick diagram to avoid an open S/D line.

Layout of basic gates has a significant influence on the occurrence of open defects. Layout of a logic gate can be modified such that the probability of a
SOP fault is reduced or eliminated. Koeppe presented a set of layout rules to
deal with SOP faults [40]. By application of these rules, SOP faults are either
reduced or their detection is simplified. He argued that the SOP faults, in general, are caused by missing contacts, cracks in metal over oxide steps, and dust
particles. For the S/D-line faults, only faults in the parallel branches (p-channel transistors in NAND gates) of a basic logic gate (NAND, NOR) require 2-
pattern test sequences. To detect such SOP faults, he suggested a reduction in
the contact locations. Fig. 2.10 illustrates the conventional and an alternative
stick diagram for a 2-input NAND gate. In the alternative stick diagram a contact is placed such that its absence affects all parallel branches together.
Therefore, such a fault causes a SA (Z/0) fault and is detected by the SA test
vector set. Similarly, for gate-line SOP faults, he suggested branchless and
fixed order routing of signals inside the logic gates such that the chance of an
open defect causing a single SOP transistor is reduced. In certain instances, it
may not be possible to reduce contact locations. In such instances, placement

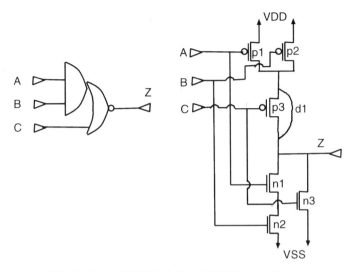

Fig. 2.11: A SON fault in a CMOS complex gate.

of an extra contact reduces probability of a fault and improves the yield, often without area penalty.

To investigate the effectiveness of the rules, Koeppe performed the fault simulation over original and modified layouts. The fault simulation results demonstrated that the SOP fault coverage of the SA test vector set increased substantially for the modified layout. The area overhead for the implementation of the rules was also low. It was expected to be between 0% and 20% depending upon the application of the rule and the original style of the cell layout. A small performance degradation was also expected since the parasitic capacitance of the transistor drains in the alternative layout was higher than in the original layout.

2.3.2.2 Transistor Stuck-On Fault Model

A stuck-on (SON) fault forces a transistor to be in the conduction mode irrespective of the voltage on its gate terminal. Fig. 2.11 illustrates a defect causing a transistor SON fault. The figure shows a 3-input AND-NOR complex gate. A bridging defect between source and drain of the p3 transistor causes a SON fault. SON faults cause state dependent degradation of Boolean output

levels. Therefore, their detection depends on circuit level parameters. In order
to detect the SON fault in transistor p3, the input test vector ABC is chosen as
001, 0X1 or X01. In the fault free case, the output should be logic low. However, due to the SON fault, a conflict is created between n3 and p-channel transistors that causes a resistive ladder between VDD and VSS. The voltage on
the output Z depends on the on-resistance of the transistor n3, resistance of the
defect d1, and the on-resistance of conducting p-channel transistors. Often
this voltage lies in the ambiguous region between logic 0 and logic 1: Hence,
such faults are very difficult to detect by logic testing.

A unique property of CMOS circuits can be exploited to test for SON faults.
The steady state current (I_{DDQ} as it is popularly known) consumption in
CMOS circuits is extremely low. A million transistor IC may have an I_{DDQ}
value of less than 100 nano amperes. As we know, a SON fault at appropriate
input logic conditions causes a resistive ladder between VDD and VSS nodes.
The steady state current flow through this resistive path will give the indication of a SON fault. For most practical situations, the difference between
faulty and the fault-free I_{DDQ} is sufficiently large for an unambiguous fault
detection [81].

Similarly, there have been efforts to detect SON faults with delay testing,
however, with limited success. It can be shown that a SON fault affecting an
n-channel transistor in a primary CMOS gate will cause an extra delay in the
0->1 output transition under certain input conditions. However, the same defect may speed up 1->0 output transition under different input conditions. In a
study over CMOS logic gates Vierhaus et al. [92] found that in general SON
faults can not be safely tested with delay testing but are effectively tested with
I_{DDQ} testing.

2.3.3 Layout Level Fault Modeling

Layout level fault modeling is motivated by several factors. First and foremost
amongst them is the inability of logic and transistor level fault models to represent physical defects with desired accuracy. Many defects (e.g., gate oxide

defects) degrade the transistor behavior that can not be mapped onto a transistor level fault model. Similarly, bridging faults in interconnects can not be mapped to higher level fault models. Higher packing densities and smaller feature sizes make such defects more likely in current technologies. In other words, a large number of potential defects can not be modeled by transistor or logic level fault models. The rising quality objective of 10 PPM or less necessitates that the layout information is exploited to generate better and effective fault models. All faults are assumed to be equally probable in logic and transistor level fault models. However, this is rarely the case in reality. Some faults are more likely than others. This information should be exploited not only for efficient, effective and economic test generation but also for creation of defect insensitive layouts. This subject will be treated at length in the subsequent pages of this chapter.

2.3.4 Function Level Fault Modeling

Over the years semiconductor technology has matured to such an extent that a VLSI chip today contains a variety of digital and analog functional blocks. The motivation behind this integration is to offer cheaper and reliable system solutions. Very often a single IC package may contain all functional blocks of an entire micro-controller or DSP processor including memories and analog interfaces. These functional blocks or system components must share a common substrate and the manufacturing process. This development has resulted in dramatic changes for testing. In spite of advances in CAD tools and CPU power, it is no longer possible to simulate faults in a complex IC at the transistor level of abstraction. Instead a complex IC is divided into many functional modules or macros [6]. In many cases, it is possible to model (or map) the impact of defects on the function of the macro. Once this mapping is known test vectors can be generated. However, this mapping must be repeated for each transistor level implementation.

Testing of semiconductor RAMs is a typical example of functional level fault modeling. However, RAM fault modeling, test algorithm development, and testing is a mature discipline by itself. A lot of attention has been paid on modeling [16,33,69] and testing of faults in RAM [57,72,74,85,87]. In Chapter 5 we address the defects and their detection strategies for RAMs. Hence, in this sub-section, we only address the function level fault modeling taking RAMs as a vehicle. Furthermore, for a tutorial overview an interested reader is referred to [1,14,29,45,62].

Thatte and Abraham [87] suggested a functional test procedure for semiconductor RAM testing. They argued that all RAM decoder and Read/Write logic faults can be mapped onto the RAM matrix as inter-cell coupling faults. An address decoder is combinational logic that selects a unique RAM cell for a given address. Assuming that under faulty conditions, the address decoder stays combinational it will behave in one of the following manners:

1. The decoder will not access the addressed cell. In addition, it may access another cell.

2. The decoder will access multiple cells, including the addressed cell.

Both of these faulty situations can be viewed as coupling faults involving two or more RAM cells. Similarly, impact of Read/Write logic faults is viewed as the SAF and/or coupling fault in the RAM matrix. On the basis of these arguments, the authors evolved efficient algorithms (complexity $n\log n$) compared to more complex (n^2) methods prevalent in the 70s. Similarly, there had been other attempts to model RAMs [16,33,57,69,72,74,85], PLAs [79] and microprocessors [8,88].

2.3.5 Delay Fault Models

In digital circuits the input-output relationship is Boolean in nature. The logic and transistor level fault models describe the steady state malfunctioning of the Boolean relationship, but can not model the faulty delay behavior of a logic element. Timing (or delay) is also an important design parameter in the input-output relationship. An otherwise good IC may fail to perform correctly in a system if it fails to meet designed timing specifications. With increasing system complexities and higher operational frequencies, timing is becoming a very important aspect of the design. Furthermore, rising quality expectations, motivates testing for the correct temporal behavior, commonly known as delay testing [10]. Generally, pre-fabrication timing is verified at each successive levels of design hierarchy. At each level the objective of the analysis is either to determine the maximum operational frequency at which circuit will behave correctly, or to guarantee that the circuit operates without any malfunction at pre-specified clock rate [3]. Once a chip is fabricated, it still must be tested for pre-specified clock frequency. A circuit is said to have a delay fault if the output of the circuit fails to reach its final value within pre-specified timing constraint.

A timing or delay fault in an IC could be caused by a number of reasons that include subtle manufacturing process defects, transistor threshold voltage shifts, increased parasitic capacitance, improper timing design, etc. A substantial research effort has been directed towards delay fault testing [10,13,36,55,58,60,61,78,90,91]. Broadly, two fault models have been proposed for delay fault testing in the literature.

2.3.5.1 Gate Delay Fault Model

Each gate in an IC is designed with a pre-specified nominal delay. However, under the gate delay fault model, the faulty gate may assume considerably larger delay. The test complexity of gate delay fault model is relatively small compared to that of the path delay fault model. This is because, in a digital IC the number of paths can be exponential in the number of gates. It appears that SA or SOP faults are special (limiting) cases of the gate delay faults. For example, in the case of a SAF the logic gate output has an infinite delay for a class of input stimuli. Similarly, for a SOP fault the transistor has infinite delay. However, there is an important distinction between SAF or SOP faults and delay faults. Unlike, SA or SOP faults, a gate delay fault does not necessarily cause the circuit to malfunction. In other words, a faulty gate may assume significantly larger delay than its nominal delay and still the circuit could work within the timing constraints. Therefore, in general, an evaluation scheme for a delay fault test must not only compute whether or not a delay fault is detected but also calculate the size of the fault. The size of a delay fault is defined as the fault detection size (FDS) [60] or as detection threshold [36] for a test, T. The FDS of a fault for the test T has the property that T is guaranteed to detect any fault at that site that is greater than the FDS. However, best FDS achievable for any gate delay fault detecting test is the corresponding slack at the fault site [36,60]. The slack of a signal is defined as the difference between clock period and the propagation delay of the longest delay path through that signal. Therefore, the quality of the test set depends how small a delay fault can be tested by a test. Methods for designing tests that activate the longest sensitizable path through every gate have been discussed in the literature [44,56].

The major drawback of the gate delay fault model is that the interconnect delay is not considered. This was acceptable when the feature size was relatively large and the gate delay was relatively large compared to the interconnect delay. However, scaling of process dimensions has changed this equilibrium. The transistor switching times have reduced dramatically due to smaller geometries. As line widths are scaled into deep sub-micron regime and device switching speed continues to improve, however, delays due to interconnects have not scaled. In fact, the interconnect resistance has increased due to smaller cross sectional area. Furthermore, there is no significant decrease in the interconnect capacitance. Therefore, interconnect RC product has increased [4,7,26,73]. Therefore, gate delay fault model is restrictive in its application to finer geometries. Furthermore, the gate delay fault model can not account for the cumulative effect of small delay variations along paths from primary inputs to primary outputs.

2.3.5.2 Path Delay Fault Model

The path delay fault model considers the cumulative delay of paths from primary inputs to primary outputs. The path delay fault model, in addition to the single isolated failures, also considers distributed delay effects due to statistical process variations. A faulty situation may arise in spite of the fact that each individual component meets its individual delay specifications. A path delay test will detect both localized as well as distributed delay defects. Path delay faults may also provide a mechanism for monitoring process variations that may have significant impact on critical paths. Furthermore, they provide an ideal vehicle for speed-sorting since they have the most accurate description of the clock speed at which timing failures begin to occur [46,80].

However, the path delay fault model has the disadvantage that it is practical to generate tests for only small number of total paths in a given circuit. Hence, the path delay fault coverage tends to be low [91]. For all practical purposes, the delays in the longest and the shortest paths (critical paths) are considered. If these delays are within the clock cycle, the circuit is considered to be delay fault-free, otherwise it contains a path delay fault.

2.3.5.3 Robust and Non-Robust Tests for Path Delay Faults

Delay fault testing assumes that delay of a gate (or a path) depends on the transition propagated from the input to the output. However, the fault is independent of the vector that induces the given transition. Testing a delay fault requires a 2-pattern tests, $<V_1, V_2>$ [10]. Similar, to the SOP testing, the first test vector, V_1, is called initializing test vector and the second vector V_2, is called as fault exciting test vector.

Irrespective of the fault mode (gate or path delay) 2-pattern test may be categorized as robust or non-robust. A robust test detects the targeted delay faults irrespective of the presence of other delay faults in the circuit. Numerous classifications of the robust path delay fault test exist [10,11,42,78], e.g., hazard free robust test, single/multiple input changing tests, and single/multiple path propagating tests. A necessary and sufficient set of path delay faults known as primitive delay faults must be tested to guarantee timing correctness of the circuit [39]. An important property of a path delay test is that the test must not be invalidated by variable delays of the fan-in signals of gates on the targeted path. On the other hand, a non-robust test detects the fault if no other delay faults affects the circuit. These faults are statically sensitizable. A fault is statically sensitizable if there exists at least one input vector which stabilizes all side inputs of the gates on the target path at non-controlling values [78]. In general, it is possible to determine whether or not a given 2-pattern test is robust by examining the logic structure of the circuit under test. The actual circuit delays are not important in this determination [58].

2.3.6 Leakage Faults

As mentioned before, static CMOS circuits have very low quiescent current (I_{DDQ}). Most manufacturing defects in CMOS ICs exhibit state dependent elevated I_{DDQ}. Therefore, I_{DDQ} testing is a powerful test method in manufacturing process defect detection. Defect-free MOS transistor has nearly infinite input impedance, hence, there should not be any current flow between gate and source, gate and drain, or gate and substrate (well). However, some defects, such as a gate oxide short, will cause leakage current flow between gate and other nodes of the transistor. In general, a leakage fault may occur between any two nodes of a MOS transistor. Nigh and Maly [54] and Mao et al. [50] independently proposed a leakage fault model containing six types of faults for MOS transistors:

- f_{GS} -- leakage fault between gate and source
- f_{GD} -- leakage fault between gate and drain
- f_{SD} -- leakage fault between source and drain
- f_{BS} -- leakage fault between bulk and source
- f_{BD} -- leakage fault between bulk and drain
- f_{BG} -- leakage fault between bulk and gate

These faults include not only the gate oxide defect causing leakage but also the leakages between various diodes required to realize a MOS transistor. Furthermore, Nigh and Maly [54] suggested that well to substrate diode defects need not be considered explicitly as leakage or latchup caused by them is easily observable.

Leakage faults such as gate oxide shorts or *pn*-junction pinholes occur quite frequently in the CMOS process. Furthermore, reduced geometries increase the electric field in MOS transistors which may cause successive degradation of gate oxide, etc. Typically, small leakage faults do not cause a catastrophic failure of an IC. However, they are potential reliability hazards [81].

2.3.7 Temporary Faults

Unlike most other faults, the temporary faults can not be classified according to levels of fault modeling. Therefore, they should be treated separately. As their name suggests, temporary faults are not permanent in nature. A major portion of digital system malfunctions is caused by temporary faults. They are harder to detect because at the time of testing they are not reproduced. There are two types of temporary faults (i) transient faults or intermittent faults and, (ii) reliability faults.

2.3.7.1 Transient or Intermittent Faults

Transient or intermittent faults are non-recurring temporary faults. Typically they are caused by alpha-particle radiation or power supply fluctuations. Transient faults can also be caused by capacitive, inductive coupled disturbances and an external electromagnetic field. They are not repairable because they do not cause physical damage to the hardware. Dynamic logic and memories are particularly susceptible to such faults

2.2.7.2 Reliability Faults

Reliability faults are often recurring in nature and appear at regular interval. These faults are caused by circuit parameter degradations, aging [53] or soft defects. This degradation is progressive until a permanent failure occurs. These faults can also occur due to design sensitivity to environmental conditions like ambient temperature, humidity, vibrations, etc. The frequency of their occurrence depends on how effective an IC (system) is protected against environmental conditions through cooling and shielding.

2.4 Inductive Fault Analysis

The circuit layout influences the impact of a defect, and thus the faulty circuit behavior to a large extent. This information is often ignored while developing fault models at transistor or logic level. In one of the earliest papers on the subject, Galiay et al. [26] pointed out that the layout information should be utilized while developing the fault model. They argued that all failures can be modeled by stuck-at-0 and stuck-at-1, will not be a sound assumption as IC density increases. In a study over 4-bit microprocessor chips they found that most failures were due to shorts and opens. Furthermore, gate level fault models do not adequately represent the faulty behavior. To test a given circuit, they suggested that an analysis of the failure mechanisms at the layout level should be carried out. In another study, Banerjee and Abraham [5] demonstrated that understanding the effects of physical failures on digital systems is essential to design tests for them and to design circuitry to detect or tolerate them. In yet another study, Maly, et al. [76] proposed a methodology of mapping physical manufacturing process defects to circuit-level faulty behavior caused by these defects. In this manner layout and technology specific faults are generated and ranked according to their likelihood. One conclusion of their analysis is that manufacturing process defects can give rise to a much broader range of faults than can be modeled using single line stuck-at fault model. Similarly, other studies point out that a lot of information can be extracted from the layout [16,21,47].

These early publications [5,26,47,75] on the subject formed the basis of what is known today as the inductive fault analysis (IFA). The IFA differentiates itself from the conventional fault modeling approaches of assuming faulty behaviors on interconnects or logic gates. It derives the circuit or logic-level fault model starting from particular physical defects. In other words, a higher-level fault model is formulated by examining defects at lower level or defects are induced by simulation of the defect creation process. Hence the word "inductive" which means the higher-level fault information is induced from lower level defects. Often IFA is referred to as realistic fault analysis and fault modeling based on IFA is referred to as realistic fault modeling. The term realistic signifies that each fault has a physical basis, i.e., a defect. A test approach based on IFA is also referred to as defect oriented testing.

In order to fully exploit the potential of IFA, it is important to understand the relationship between manufacturing defects and IC faults. For example, many defects that influence the IC performance are tested before the functional testing. Therefore, they are not included in IFA based test generation. Similarly, manufacturing defects in an IC and their impact on performance is strongly influenced by the IC design and layout. IFA can also be exploited to find out areas in the design that are difficult to test. Using this information the design robustness and yield can be improved. In the following sub-sections, an overview of defect-fault relationship and IC design and layout related sensitivity will be given.

2.4.1 Defect-Fault Relationship

Although the process yield has improved dramatically over the years, the ever increasing quest for more functions on a single IC has led to the shrinkage of device geometries and increase in chip area. Unfortunately, both of these developments have caused ICs to become susceptible to various yield loss mechanisms. In final terms, it is the yield of an IC that determines whether or not greater integration is an economically good proposition. Hence, it has become increasingly relevant to know different yield loss mechanisms. The IC manufacturing process involves a sequence of basic processing steps performed on a batch of wafers. Maly et al. [49] described that the outcome of a manufacturing operation depends on three major factors: the process controlling parameters or *control*, the *layout* of the IC, and some randomly changing environmental factors, called *disturbances*. The control of a manufacturing operation is the set of parameters that should be manipulated for desired changes in the fabricated IC structure. The layout of an IC is the set of masks distinguishing IC areas which need to be processed for each manufacturing step. The disturbances are environmental factors that influence the result of the manufacturing operation. These manufacturing process disturbances have been studied in great detail [27,49,86] and are classified as:

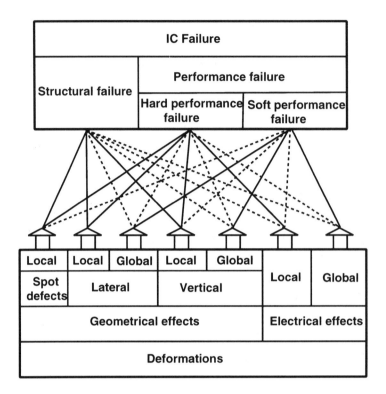

Fig. 2.12: IC manufacturing process deformation and their relationship
with IC faults [49].

- Human errors and equipment failures

- Instabilities in the process conditions

- Material instabilities

- Substrate inhomogeneities

- Lithography spots

A detailed treatment of manufacturing process disturbances can be found in
the above mentioned references. Most disturbances influence the processed
topology of the IC. However, for the purpose of realistic (or IFA based) fault
modeling it is important to know that not all disturbances influence the IC performance equally. In other words, these disturbances deform the IC and hence

can be grouped according to classes of deformations. A disturbance is the phenomenon that leads to a deformation in an IC. For example, a contamination (disturbance) on the wafer causes a break (deformation) in the metal line. In this case the deformation is geometrical in nature. Similarly, a poor temperature control (disturbance) during the growth of gate oxide may results in lower threshold voltage (electrical deformation). In general, all process disturbances can be classified into geometrical and electrical deformations [49].

Fig. 2.12 shows this classification and its relationship with different IC failures. The lower half of the figure shows the classification of physical phenomena that cause yield loss and the upper half of the figure shows the fault classification (structural and performance faults). Geometrical and electrical deformations have local as well as global influences on IC function and/or performance. A global influence occurs when a particular parameter, say the transistor threshold, is affected over the complete wafer. The term local is used when the influence on the parameter is limited to a region smaller than a wafer. Often these local deformations are called defects like break and short in conductors. In addition, spot defects that are primarily lithographic in nature form a part of geometrical deformations. In principle, each class of physical deformation is capable of causing a variety of faults. However, some are more likely (solid lines) than others. For example, all global effects are more likely to cause soft performance failures. Similarly, spot defects are more likely to cause structural or hard performance failures. Since, the impact of global deformations affects the complete or a large part of a wafer, they are quickly detected by test structures designed for them. Furthermore, in a well controlled fabrication environment such problems are kept under control. Therefore, for IFA based fault modeling and testing, only the local deformations or defects are taken into account.

2.4.2 IC Design and Layout Related Defect Sensitivity

The design of a modern IC is a complex task. Often designs tend to exploit the very maximum of what a manufacturing process can offer. Issues like time to market, time to profitability, and reduced product life cycle further complicate

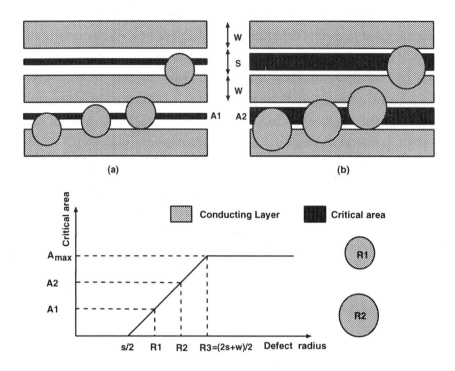

Fig. 2.13: Critical area as the function of defect radius.

the decisions in the design. Higher operational frequency, higher complexity, smaller area, lower power consumption, usually are the design objectives. Often these objectives are in conflict with each other and are rarely in agreement with what is known as robust design practices. Typically, it takes a lot of product engineering and a couple of design iterations to stabilize the design for reasonably good yield. In spite of all this effort, it has been observed that yields of certain designs, for a given chip area, are lower compared to those of others. Design related sensitivity to yield can be divided into two major classes that are further divided into subclasses.

2.4.2.1 Defect sensitive design

The operational frequency of a digital IC is determined by its critical path. A critical path is a data path in an IC with the largest delay. For a correct operation of the chip, the critical path delay should be less than the clock period.

However, the actual delay of the critical path is governed by the process. The design margin between the critical path and the clock frequency should be reasonable, otherwise parametric process variations or spot defects may result into timing (parametric) failures. Furthermore, how much physical area does the critical path have on the chip is also an important issue because higher area will increase the probability of a defect landing onto the critical path. An otherwise innocuous defect in the critical path may increase its parasitic capacitance and/or resistance resulting in a timing failure. Similarly, the logic implementation also has an influence over the timing related sensitivity of the design. For example, the timing critical design aspects are much higher in the dynamic logic implementation compared to the static logic implementation. Furthermore, in the dynamic logic implementations, often logic levels are defined under the dynamic conditions. Many defects influence the dynamic behavior and hence cause failures. On the other hand, impact of such defects on the static logic is not so severe, but they cause increased delay. If this delay is not in the critical path, it may not lead to a failure. Moreover, the type of synchronous logic implementation, random logic, on board memories and PLA, all have yield related repercussions because they have different inherent sensitivities to defects.

2.4.2.2 Defect sensitive layout

As mentioned earlier, the IC layout also contains a wealth of yield related information. Researchers [22,48] have defined the concept of *critical area*. This area is defined for a defect of radius R, as the area on the die in which the center of a circular defect must fall for a fault to occur in the circuit. It is illustrated in Fig. 2.13. For a given density distribution of a particular defect, one can find out the most probable defect radius. Now, for a given defect radius of a particular defect, the critical area can be computed. It can be deduced that higher the critical area in a given IC, the lower will be the yield. This knowledge can be used to minimize the critical area. Thus, the placement and routing strategies and their implementation have an impact on the yield. IFA can be exploited to find out the layout or design related yield sensitivity. Moreover, by this methodology better design and layout can be evolved. Similarly, it can be argued that clocking strategies including clock routing and various possible implementations will have impact on the IC yield.

2.4.3 Basic Concepts of IFA

IFA is a systematic approach for determining what faults are likely to occur in a VLSI circuit. This approach takes into account the technology of the implementation, the circuit topology and the defect statistics of the fabrication plant. Shen et al. [76] formalized concepts of IFA as follows: *IFA is a systematic procedure to predict all the faults that are likely to occur in a MOS integrated circuit or sub-circuit. The three major steps of the IFA procedure are; (1) generation of physical defects using statistical data from fabrication process, (2) extraction of circuit level faults caused by these defects, and (3) classification of fault types and ranking of faults based on their likelihood of occurrence.*

The major steps of the IFA are shown in Fig. 2.14. The circuit layout and the manufacturing defect statistics form inputs to the analysis. The IFA methodology takes into account only local deformations or defects. The size and the probability of defects is defined by the defect density distribution (DDD). For an IC manufacturing fab, DDDs are normally available for each layer and defect types. DDDs define how large the probability of a defect in a certain layer is with respect to other layers, and how the probability of defects in a layer depends on the size of the defect. For a typical double metal single poly CMOS process, these defects include:

- Extra and missing material defects in conducting and semiconducting layers

- Presence of extra contacts and vias

- Absence of contacts and vias

- Thin and thick oxide pinholes

- Junction leakage pinholes

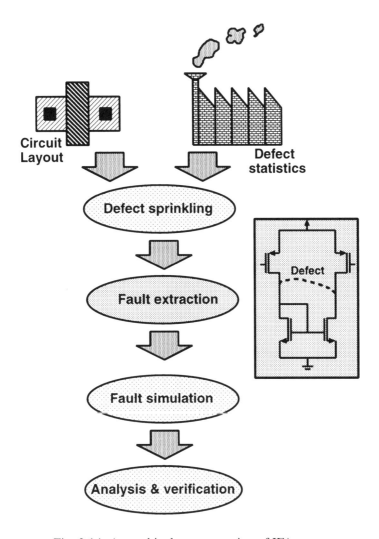

Fig. 2.14: A graphical representation of IFA.

The defects are sprinkled onto the layout in a random manner. For this purpose, a CAD tool like VLASIC (Vlsi LAyout Simulation for Integrated Circuits) [93,94] or DEFAM (DEfect to FAult Mapper) [25] is utilized. The defects are modeled as absence or presence of material on the layout. Only a small subset of all defects cause a change in the circuit connectivity. For ex-

ample, a short is created between two nodes or an open defect causes a break
in the connectivity. These defects are extracted and their impact on circuit behavior is modeled at an appropriate level of abstraction for fault simulation.
Subsequently, the abstracted defects (i.e., faults) are simulated with given test
stimuli. The fault simulation information is exploited for providing DfT solutions, building fault tolerance into the circuit, etc. Finally, simulation results
are verified by the silicon data.

The IFA technique takes into account only a subset of process defects, namely
local defects (Fig. 2.12). The global defects or deviations cause widespread
failures or near failures. Since, the impact of global defects is present over a
relatively large area, these are detected rather quickly. Moreover, there are
special test structures to test for these erroneous conditions. Hence, such defects are detected early in the manufacturing process before the functional/
structural testing. Furthermore, in a well controlled and monitored process,
major process errors causing global defects are relatively easily detected and
solved.

2.5 Conclusion

The functional testing of complex digital ICs is prohibitively expensive and
does not ensure that the IC is fault-free. Structural tests that target faulty circuit behavior provide an alternative. The effectiveness of a structural test is
quantifiable in terms of the covered faults. Thus, it allows the user to establish
a relationship between the test coverage and the quality of the tested devices.
The test generation for structural test is considerably simpler due to availability of CAD tools. However, the structural test requires a fault model that represents likely manufacturing process defects with an acceptable accuracy and
provides an objective basis for the structural test generation. A number of fault
models are available and most of them are classified according to the level of
abstraction. Gate level (SAF), transistor level (SOP, SON) and function level
are some of the examples of the abstraction levels. The level of abstraction is
essentially a compromise between the fault model's ability to represent actual

defects and the speed of processing of the fault in a fault simulation environment.

The conventional fault modeling approaches do not consider the likely or realistic faults in a given layout of a circuit. The layout of a circuit has a significant impact on the faulty circuit behavior. The IFA takes into account the circuit layout and the defect data from the manufacturing site to generate a list of realistic faults. The word realistic signifies that each fault has a physical basis (i.e., defect). In this manner the circuit layout dependent fault models are evolved. Many reported experiments illustrate the effectiveness of the method in realistic fault model generation with success.

References

1. M.S. Abadir, and H.K. Reghbati, "Functional Testing of Semiconductor Random Access Memories," *ACM Computing Surveys*, 15(3), pp. 175-198, Sept. 1983.

2. J. A. Abraham, "Fault Modeling in VLSI," in *VLSI Testing, Edited by T.W. Williams*, vol. 5, North-Holland, pp. 1-27, 1986.

3. V.D. Agrawal, "Synchronous Path Analysis in MOS Circuit Simulator," *Proc. of 19th Design Automation Conf.*, June 1982, pp. 629-635.

4. H.B. Bakoglu and J.D. Meindel, "Optimal Interconnection Circuits for VLSI," *IEEE Transactions on Electron Devices*, vol. ED-32, no. 5, pp. 903-909, May 1985.

5. P. Banerjee and J.A. Abraham, "Characterization and Testing of Physical Failures in MOS Logic Circuits," *IEEE Design & Test of Computers*, vol. 1, pp. 76-86, August 1984.

6. F.P.M. Beenker, K.J.E. van Eerdewijk, R.B.W. Gerritsen, F.N. Peacock and M. van der Star, "Macro Testing, Unifying IC and Board Test," *IEEE Design and Test of Computers*, vol. 3, pp. 26-32, December 1986.

7. S. Bothra, B. Rogers, M. kellem and C.M. Osburn, "Analysis of the Effects of Scaling on Interconnect Delay in ULSI Circuits," *IEEE Transactions on Electron Devices*, vol. ED-40, no. 3, pp. 591-597, March 1993.

8. D.S. Brahme and J.A. Abraham, "Functional Testing of Microprocessors," *IEEE Transactions on Computers*, vol. C-33, pp. 475-485, 1984.

9. M.A. Breuer and A.D. Friedman, *Diagnosis and Reliable Design of Digital Systems,* Woodland Hills, California: Computer Science Press, 1976.

10. T.J. Chakraborty, V.D. Agrawal and M.L. Bushnell, "Delay Fault Models and Test Generation for Random Logic Sequential Circuits," *Proc. 29th Design Automation Conf.*, June 1992, pp. 165-172.

11. R. Chandramouli, "On Testing Stuck-Open Faults," *Proceedings of 13th Annual International Symposium on Fault Tolerant Computing Systems*, 1983, pp. 258-265.

12. K.T. Chang and H.C. Chen, "Classification and Identification of Nonrobust Untestabel path Delay Faults," *IEEE Transactions on CAD,* vol. 15, pp. 845-853, August 1996.

13. K.T. Cheng, "Transition Fault Simulation for Sequential Circuits," *Proceedings of International Test Conference*, 1992, pp. 723-731.

14. B.F. Cockburn, "Tutorial on Semiconductor Memory Testing," *Journal of Electronic Testing: Theory and Applications (JETTA),* vol. 5, no. 4, pp. 321-336, November 1994.

15. H. Cox and J. Rajaski, "Stuck-Open and Transition Fault Testing in CMOS Complex Gates," *Proceedings of International Test Conference*, 1988, pp. 688-694.

16. R. Dekker, F. Beenker and L. Thijssen, "Fault modeling and Test Algorithm Development for Static Random Access Memories," *Proceedings of International Test Conference,* 1988, pp. 343-352.

17. C. Di and J.A.G. Jess, "On Accurate Modeling and Efficient Simulation of CMOS Open Faults," *Proceedings of International Test Conference,* 1993, pp. 875-882.

18. E.B. Eichelberger and T.W. Williams, "A Logic Design Structure for LSI Testability," *Journal of Design Automation and Fault Tolerant Computing*, vol. 2, no. 2, pp. 165-178, May 1978.

19. R.D. Eldred, "Test Routines Based on Symbolic Logical Statements," *Journal of ACM*, vol. 6, no.1, pp. 33-36, January 1959.

20. Y.M. El-Ziq and R.J. Cloutier, "Functional-Level Test Generation for Stuck-Open Faults in CMOS VLSI," *Proceedings of International Test Conference*, 1981, pp. 536-546.

21. F.J. Ferguson and J.P. Shen, "Extraction and Simulation of Realistic CMOS Faults using Inductive Fault Analysis," *Proceedings of International Test Conference*, 1988, pp. 475-484.

22. A.V. Ferris-Prabhu, *Introduction to Semiconductor Device Yield Modeling,* Boston: Artech House, 1992.

23. M.L. Flottes, C. Landrault and S. Pravossoudovitch, "Fault Modeling and Fault Equivalence in CMOS Technology," *Journal of Electronic Testing: Theory and Applications,* vol. 2, no.3, pp. 229-241, August 1991.

24. S. Funatsu, N. Wakatsuki and T. Arima, "Test Generation Systems in Japan," *Proceedings of 12th Design Automation Symposium*, 1975, pp. 114-122.

25. D. Gaitonde and D.H.H. Walker, "Test Quality and Yield Analysis Using the DEFAM Defect to Fault Mapper," *Proceedings of International Conference on Computer Aided Design*, 1993, pp. 202-205.

26. J. Galiay, Y. Crouzet and M. Vergniault, "Physical Versus Logical Fault Models in MOS LSI Circuits: Impact on Their Testability," *IEEE Transaction on Computers*, vol. C-29, no. 6, pp. 527-531, June 1980.

27. S.K. Gandhi, *VLSI Fabrication Principles,* John Wiley and Sons, 1983.

28. D.S. Gardner, J.D. Meindel and K.C. Saraswat, "Interconnection and Electromigration Scaling Theory," *IEEE Transactions on Electron Devices,* vol. ED-34, no. 3, pp. 633-643, March 1987.

29. A.J. van de Goor, *Testing Semiconductor Memories: Theory and Practices,* John Wiley and Sons, 1991.

30. A. Goundan and J.P. Hayes, "Identification of Equivalent Faults in Logic Networks," *IEEE Transactions on Computers*, vol. c-29, no. 11, pp. 978-985, November 1980.

31. J.P. Hayes, "Fault Modeling for Digital Integrated Circuits," *IEEE Transactions on Computer-Aided Design of Circuits and Systems*, CAD-3, pp. 200-207, 1984.

32. J.P. Hayes, "Fault Modeling," *IEEE Design & Test of Computers*, vol. 2, pp. 88-95, April 1985.

33. J.P. Hayes, "Detection of Pattern-Sensitive Faults in Random Access Memories," *IEEE Transactions on Computers*, vol. C-24, no.2, pp. 150-157, February 1975.

34. R.J.A. Harvey, A.M.D. Richardson, E.M.J. Bruls and K. Baker, "Analogue Fault Simulation Based on Layout Dependent Fault Models," *Proceedings of International Test Conference*, 1994, pp. 641-649.

35. O.H. Ibarra and S.K. Sahni, "Polynomial Complete Fault Detection Problems," *IEEE Transactions on Computers*, vol. c-24, no. 3, pp. 242-249, March 1975.

36. V.S. Iyenger et al., "On Computing the Sizes of Detected Delay Faults," *IEEE Transactions on CAD*, vol. 9, no. 3, 299-312, 1990.

37. S.K. Jain and V.D. Agrawal, "Modeling and Test Generation Algorithm for MOS Circuits," *IEEE Transactions on Computers,* vol. 34, no. 5, pp. 426-43, May 1985.

38. A.P. Jayasumana, Y.K. Malaiya and R. Rajsuman, "Design of CMOS Circuits for Stuck-Open Fault Testability," *IEEE Journal of Solid-State Circuits*, vol. 26, no. 1, pp. 58-61, January 1991.

39. W. Ke and P.R. Menon, "Synthesis of Delay Verifiable Combinational Circuits," *IEEE Transactions on Computers*, vol. 44, pp. 213-222, February 1995.

40. S. Koeppe, "Optimum Layout to Avoid CMOS Stuk-Open Fault," *Proceedings of 24th ACM/IEEE Design Automation Conference,* 1987, pp. 829-835.

41. F.C.M. Kuijstermans, M. Sachdev and L. Thijssen, "Defect Oriented Test Methodology for Complex Mixed-Signal Circuits," *Proceedings of European Design and Test Conference*, 1995, pp. 18-23.

42. W.K. Lam, A. Saldanha, R.K. Brayton and A.L. Sangiovanni-Vincentelli, "Delay Fault Coverage and Performance Trade-offs," *Proceedings of 30th Design Automation Conference*, 1993, pp. 446-452.

43. K.J. Lee and M.A. Breuer, "On the Charge Sharing Problem in CMOS Stuck-Open Fault Testing," *Proceedings of International Test Conference*, 1990, pp. 417-425.

44. A.K. Majhi, J. Jacob, L.M. Patnaik and V.D. Agrawal, "On Test Coverage of Path Delay Faults," *Proc. of 9th International Conference on VLSI Design*, January 1996, pp. 418-421.

45. P. Mazumder and K. Chakraborty, *Testing and Testable Design of High-Density Random-Access Memories,* Boston: Kluwer Academic Publishers, 1996.

46. Y.K. Malaiya and R. Narayanaswamy, "Modeling and Testing for Timing Faults in Synchronous Sequential Circuits," *IEEE Design & Test of Computers*, vol. 1, no. 4, pp. 62-74, November 1984.

47. W. Maly, F.J. Ferguson and J.P. Shen, "Systematic Characterization of Physical Defects for Fault Analysis of MOS IC Cells," *Proceedings of International Test Conference*, 1984, 390-399.

48. W. Maly, W.R. Moore and A.J. Strojwas, "Yield Loss Mechanisms and Defect Tolerance," SRC-CMU Research Center for Computer Aided Design, Dept. of Electrical and Computer Engineering, Carnegie Mellon University, Pittsburgh, PA 15213.

49. W. Maly, A.J. Strojwas and S.W. Director, "VLSI Yield Prediction and Estimation: A Unified Framework," *IEEE Transaction on Computer Aided Design*, vol. CAD-5, no. 1, pp. 114-130, January 1986.

50. W. Mao, R. Gulati, D.K. Goel and M.D. Ciletti, "QUIETEST: A Quiescent Current Testing Methodology for Detecting Leakage Faults," *Proceedings of International Conference on CAD*, 1990, pp. 280-283.

51. E.J. McCluskey and F.W. Clegg, "Fault Equivalence in Combinational Logic Networks," *IEEE Transactions on Computers,* vol. c-20, no. 11, pp. 1286-1293, November 1971.

52. A. Meixner and W. Maly, "Fault Modeling for the Testing of Mixed Integrated Circuits," *Proceedings of International Test Conference,* 1991, pp. 564-572.

53. S. Mourad and E.J. McCluskey, "Fault Models," *Testing and Diagnosis of VLSI and ULSI*, Boston: Kluwer Academic Publishers, pp. 49-68, 1989.

54. P. Nigh and W. Maly, "Test Generation for Current Testing," *IEEE Design & Test of Computers*, pp. 26-38, February 1990.

55. E.S. Park, B. Underwood, T.W. Williams and M.R. Mercer, "Delay Testing Quality in Timing-Optimized Designs," *Proceedings of International Test Conference,* 1991, pp. 879-905.

56. E.S. Park and M.R. Mercer, "An Efficient Delay Test Generation System for Combinational Logic Circuits," *IEEE Transactions on CAD,* vol. 11, pp. 926-938, July 1992.

57. C.A. Papachristou and N.B. Sahgal, "An Improved Method for Detecting Functional Faults in Semiconductor Random Access Memories," *IEEE Transactions on Computers,* vol. C-34, no.2, pp. 110-116, February 1985.

58. A. Pierzynska and S. Pilarski, "Non-Robust versus Robust," *Proceedings of International Test Conference,* 1995, pp. 123-131.

59. J.F. Poage, "Derivation of Optimum Tests to Detect Faults in Combinational Circuits," *Proceedings of Symposium on Mathematical Theory of Automata,* 1963, pp. 483-528,

60. A.K. Pramanick and S.M. Reddy, "On the Computation of the Ranges of Detected Delay Fault Sizes," *IEEE International Conference on CAD,* pp. 126-129, 1989.

61. I. Pramanick and A.K. Pramanick, "Parallel Delay Fault Coverage and Test Quality Evaluation," *Proceedings of International Test Conference,* 1995, pp. 113-122.

62. B. Prince, *Semiconductor Memories,* John Wiley and Sons, 1991.

63. R. Rajsuman, A.P. Jayasumana and Y.K. Malaiya, "CMOS Stuck-Open Fault Detection Using Single Test Patterns," *Proceedings of ACM/IEEE Design Automation Conference,* 1989, pp. 714-717.

64. R. Rajsuman, A.P. Jayasumana and Y.K. Malaiya, "CMOS Open-Fault Detection in the Presence of Glitches and Timing Skews," *IEEE Journal of Solid-State Circuits,* vol. 24, no. 4, pp. 1129-1136, August 1989.

65. S.M. Reddy and S. Kundu, "Fault Detection and Design For Testability of CMOS Logic Circuits," *Testing and Diagnosis of VLSI and ULSI,* Edited by F. Lombardi and M. Sami, pp. 69-91, 1989.

66. S.M. Reddy, M.K. Reddy and J.G. Kuhl, "On Testable Design for CMOS Logic Circuits," *Proceedings of International Test Conference,* 1983, 435-445.

67. S.M. Reddy, M.K. Reddy and V.D. Agrawal, "Robust Test for Stuck-Open Faults in CMOS Combinational Logic Circuits," *Proceedings of 14th International Symposium on Fault Tolerant Computing*, 1984, pp. 44-49.

68. B.K. Roy, "Diagnosis and Fault Equivalence in Combinational Circuits," *IEEE Transactions on Computers,* vol. c-23, no. 9, pp. 955-963, September 1974.

69. M. Sachdev and M. Verstraelen, "Development of a Fault Model and Test Algorithms for Embedded DRAMs," *Proceedings of International Test Conference,* 1993, pp. 815-824.

70. M. Sachdev, "Defect Oriented Analog Testing: Strengths and Weaknesses," *Proceedings of 20th European Solid State Circuits Conference*, 1994, pp. 224-227.

71. M. Sachdev, "A Defect Oriented Testability Methodology for Analog Circuits," *Journal of Electronic Testing: Theory and Applications*, vol. 6, pp. 265-276, June 1995.

72. M. Sachdev, "Reducing the CMOS RAM Test Complexity with I_{DDQ} and Voltage Testing," *Journal of Electronic Testing: Theory and Applications (JETTA)*, vol. 6, no. 2, pp. 191-202, April 1995.

73. K.C. Saraswat and F. Mohammadi, "Effect of Scaling of Interconnections on the Time Delay of VLSI Circuits," *IEEE Transactions on Electron Devices*, vol. ED-29, no. 4, pp. 645-650, April 1982.

74. J. Savir, W.H. McAnney and S.R. Vecchio, "Testing for Coupled Cells in Random Access Memories," *Proceedings of International Test Conference*, 1989, pp. 439-451.

75. D.R. Schertz and G. Metze, "A New Representation for Faults in Combinational Digital Circuits," *IEEE Transactions on Computers,* vol. c-21, no. 8, pp. 858-866, August 1972.

76. J.P. Shen, W. Maly and F.J. Ferguson, "Inductive Fault Analysis of MOS Integrated Circuits," *IEEE Design & Test of Computers*, vol. 2, pp. 13-26, December 1985.

77. H.C. Shih and J.A. Abraham, "Fault Collapsing Techniques for MOS VLSI Circuits," *Proceedings of Fault Tolerant Computing Symposium*, 1986, pp. 370-375.

78. M. Sivaraman and A.J. Strojwas, "Test Vector Generation for Parametric Path Delay Faults," *Proceedings of International Test Conference*, 1995, pp. 132-138.

79. J.E. Smith, "Detection of Faults in Programmable Logic Arrays," *IEEE Transactions on Computers*, vol. C-28, pp. 845-853, 1979.

80. G.L. Smith, "Model for Delay Faults Based upon Paths," *Proceedings of International Test Conference,* 1985, pp. 342-349.

81. J.M. Soden, C.F. Hawkins, R.K. Gulati and W. Mao, "I_{DDQ} Testing: A Review," *Journal of Electronic Testing: Theory and Applications*, vol. 3, pp. 291-303, November 1992.

82. M. Soma, "An Experimental Approach to Analog Fault Models," *Proceedings of Custom Integrated Circuits Conference*, 1991, pp. 13.6.1-13.6.4.

83. M. Soma, "A Design for Test Methodology for Active Analog Filters," *Proceedings of International Test Conference,* 1990, pp. 183-192.

84. M. Soma, "Fault Modeling and Test Generation for Sample and Hold Circuit," *Proceedings of International Symposium on Circuits and Systems*, 1991, pp. 2072-2075.

85. D.S. Suk and S.M. Reddy, "A March Test for Functional Faults in Semiconductor Random Access Memories," *IEEE Transactions on Computers*, vol. C-30, no.12, pp. 982-985, Dec. 1981.

86. S.M. Sze, *VLSI Technology,* New York: McGraw Hill Book Company 1983.

87. S.M. Thatte and J.A. Abraham, "Testing of Semiconductor Random Access Memories," *Proceedings of International Conference on Fault Tolerant Computing*, 1977, pp. 81-87.

88. S.M. Thatte and J.A. Abraham, "Test Generation for Microprocessors," *IEEE Transactions on Computers*, vol. C-29, pp. 429-441, 1980.

89. K. To, "Fault Folding for Irredundant and Redundant Combinational Circuits," *IEEE Transactions on Computers*, vol. C-22, no. 11, pp. 1008-1015, November 1973.

90. B. Underwood, W.O. Law, S. Kang and H. Konuk, "Fastpath: A Path-delay Test Generator for Standard Scan Designs," *Proceedings of International Test Conference*, 1994, pp. 154-163.

91. P. Varma, "On Path Delay testing in a Standard Scan Environment," *Proceedings of International Test Conference*, 1994, pp. 164-173.

92. H.T. Vierhaus, W. Meyer and U. Glaser, "CMOS Bridges and Resistive Faults: I_{DDQ} versus Delay Effects," *Proceedings of International Test Conference*, 1993, pp. 83-91.

93. SH. Walker and S.W. Director, "VLASIC: A Catastrophic Fault Yield Simulator Integrated Circuits," *IEEE Transactions on Computer Aided Design of Integrated Circuits and Systems*, vol. CAD-5, pp. 541-556, October 1986.

94. H. Walker, "VLASIC System User Manual Release 1.3," SRC-CMU Research Center for Computer Aided Design, Dept. of Electrical and Computer Engineering, Carnegie Mellon University, Pittsburgh, PA 15213.

95. R.L. Wadsack, "Fault Modeling and Logic Simulation of CMOS and MOS Integrated Circuits," *Bell Systems Technical Journal*, vol. 57, no.5, pp. 1449-1474, May-June 1978.

96. T.W. Williams and K.P. Parker, "Design for Testability--A Survey," *Proceedings of the IEEE*, vol. 71, no. 1, pp. 98-113, January 1983.

97. B.W. Woodhall, B.D. Newman and A.G. Sammuli, "Empirical Results on Undetected CMOS Stuck-Open Failures," *Proceedings of International Test Conference*, 1987, pp. 166-170.

CHAPTER 3

Defects in Logic Circuits and Their Test Implications

A substantial amount of research has been carried out since 80s to verify the validity of various fault models. This chapter summarizes some of the earlier work done in this area. During the same time frame, quiescent current measurement technique also known as I_{DDQ} testing became popular owing to its ability to uncover defects in CMOS circuits. Studies were conducted over the relative effectiveness of Boolean (logic) and I_{DDQ} methods of testing for defect detection. Salient features of these studies are reproduced.

3.1 Introduction

In the previous chapter, we reviewed various fault models and introduced the basics of defect oriented test methodology (IFA). In this chapter we shall discuss important defect oriented studies conducted on standard logic circuits. The focus is on the practical side of the concepts discussed in the previous chapter. This material provides a complementary treatment for the theoretical concepts presented in the previous chapter.

Logic circuits include combinational logic gates (INVERTER, NAND, NOR, etc.) as well as sequential circuits (flip-flops, scan chains, etc.). Standard cells

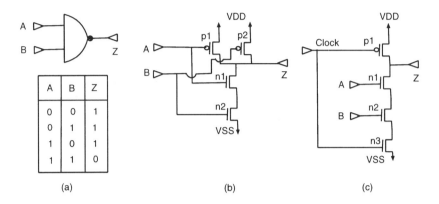

A	B	Z
0	0	1
0	1	1
1	0	1
1	1	0

(a)   (b)   (c)

Fig. 3.1: A two input NAND gate and its static CMOS (b), and dynamic
CMOS (c), implementations.

are basic building blocks used to implement a logic function. Owing to the
simplicity of standard cell logic gates, the early experiments with IFA were
conducted over standard cells. Since the SAF model had gained wide acceptance for testing and test generation, researchers typically wanted to verify
from IFA studies, "Do stuck-at faults represent manufacturing defects [4]?"

As we already know, the complementary metal oxide semiconductor (CMOS)
technology is the most popular technology today. Logic functions in CMOS
technology can be implemented with static logic gates or dynamic logic gates.
Fig. 3.1 illustrates implementation of a 2-input NAND gate in static as well as
dynamic CMOS configurations. In the static implementation (Fig. 3.1(b)), the
output Z of the logic gate retains its state so long as the inputs are unchanged.
However, in dynamic CMOS (Fig. 3.1(c)), the output Z of the logic gate has
the correct logic value at certain instance of the clock. When the clock is low
the output is precharged to logic 1 (precharge phase). At the instance clock
goes high, the output response is evaluated depending upon the states of inputs
(evaluation phase). The correct output is only available in the evaluation
phase. In the case of a dynamic NAND, if both inputs are high, the output
makes a transition to logic 0, otherwise it retains the precharged value of 1.
However, if output stays high it is not driven by any source and its value is due
to the precharged output capacitance. Very high input impedance of MOS

transistors helps dynamic logic to retain the precharged value under appropriate input conditions.

Most of the logic circuits are implemented in static CMOS technology owing to its better noise margin, robust implementation, and low power consumption. In the static CMOS implementation, a combinational logic gate has two distinct and equal parts: PMOS transistor(s), and NMOS transistor(s). The number of PMOS transistors is equal to the number of NMOS transistors. As depicted in Fig. 3.1(b) PMOS and NMOS transistors are arranged in a complementary manner with respect to each other. For example, if all PMOS transistors are in parallel, then all NMOS transistors are in series (NAND gate). Following the same principle complex logic gates can be formed such as AND-OR-INVERT, etc.

3.2 Stuck-at Faults and Manufacturing Defects

In Chapter 2 (Sections 2.2.1.3 and 2.3.1.4), we briefly touched upon the shortcomings of the stuck-at fault model in representing shorts and open defects in CMOS circuits. In this section we expand upon defects and their detection strategies in standard cell logic gates. Understanding the effect of physical failures on digital systems is essential to design test for them and to design circuitry to detect and tolerate them [3].

3.2.1 Study by Galiay, Crouzet, and Vergniult

Galiay et al. [9] studied the physical origin of failures. Their study was concluded over an NMOS 4-bit microprocessor chip. They performed failure analysis of 43 failed devices. Table 3.1 illustrates the observed failure modes of the IC. It is clear from the table that opens and shorts in metalization and diffusion were the primary causes of failures. These failures were easily detected and causes of the failures were established. A total of 10% failures did result in logical faults that were detected. However, no conclusive failure mode could be established. Similarly, 15% of failures were caused by large

Table 3.1: Failure modes observed by Galiay et al. [3].

Short between metalizations	39%
Open metalization	14%
Short between difusions	14%
Open diffusions	6%
Short between metalization and substrate	2%
Unobservable	10%
Insignificant	15%

scale imperfection like scratch, etc., that were easily detected and were considered insignificant for test purposes.

Fig. 3.2 depicts some observed failures in an NMOS logic gate identified by Galiay et al. These failures identify two broad issues in the modeling of such failures.

1. All failures can not be modeled by stuck-at faults. For example in Fig. 3.2(a), short number 1 and open number 3 can, respectively, be modeled by SA1 at input e and by SA0 at input e (or input f or both). On the other hand, short number 2 and open number 4 can not be modeled by any SAF t because they involve a modification of the function realized by the gate. For the same reason, the short between outputs of logic gates can not be modeled by SAF.

2. The actual topology of the circuit is often different from the logical representation of the circuit. Some connections in the logic circuit do not map onto the actual topology, and vice versa. Fig. 3.2(b) illustrates the electrical and logic diagrams of a complex gate. For example, short number 2 which is physically possible cannot be represented on the logic diagram. Similarly, short 1 in logic diagram has no physical meaning.

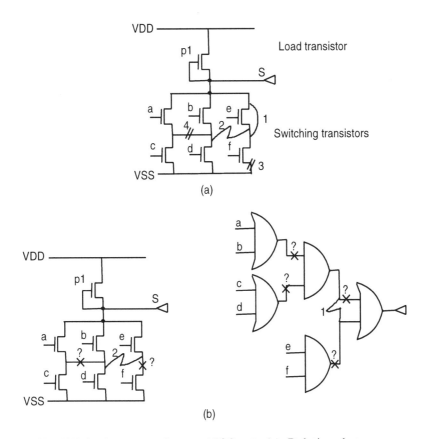

Fig. 3.2: Failure examples in a MOS gate (a); Relations between electrical and logic diagrams (b) [3].

In order to detect failures at the logic level, Galiay et al. defined the notion of conduction path as a path between output and VSS if all transistors in the path are in the conduction mode. In NMOS technology, if no conduction path exists, then the load transistor pulls the output (S) to logic high. An open defect in this switch-like network may remove one or more conduction paths. Detection of open defects in NMOS technology is simpler compared to open defect detection in CMOS [27]. Readers will recall (Chapter 2) that CMOS open defects, in general, require two test vectors, T1 and T2, for detection. In the NMOS technology, the load transistor always provides the first initializing vector (T1) for the detection of open defects in the switching network. In order to detect a particular open defect, a conduction path should be uniquely activated. The logic gate contains the above mentioned open defect if the output of the gate remains at logic high displaying a SA1 fault.

A short between two nodes inside a logic gate creates one or more conduction paths in the gate. Two conditions are required to detect a short between nodes i and j: (1) activation of at least one conduction path between i (j) and output, and activation of at least one conduction path between j (i) and VSS; and (2) blocking of all conduction paths in the logic gate. Owing to this short the output will be pulled down to logic low which is different to the output in fault-free condition. In general, there may be more than one test vectors that will detect a given short. However, this detection holds only for low resistive shorts. In the case of high resistive shorts the output level of the gate may not degrade enough so as to be interpreted as a fault by the subsequent logic gate.

3.2.2 Study by Banerjee and Abraham

Transistor level fault models represent physical failures better than the logic level fault models. Several test generation algorithms have been reported for detecting transistor level faults for combinational circuits [6-8,14,15]. Typically, transistor structure is converted into an equivalent logic gate structure. Shorts and opens were modeled as stuck-at faults at logic gate level. Although such a scheme is fairly successful in modeling most of the transistor defects, there are some defects that are not modeled. For example, a short between source and drain of transistor can not be modeled as transistor level SA fault [2,3].

Modeling of MOS circuits as a network of simple transistors (switches) began in late 70s and early 80s. Bryant presented an overview of switch level modeling and test pattern generation [5]. Switch level modeling offers several advantages. Switch level model is a close representation of schematic. Furthermore, it also models many important phenomena associated with MOS circuits, such as bidirectionality of signal flow, dynamic charge storage, resistance ratios, etc. In some sense, switch-level modeling and simulation is an excellent trade-off between accuracy of circuit level and speed of logic level modeling and simulation.

Banerjee and Abraham [3] characterized physical failures of simple NMOS and CMOS circuits and translated into logic level faults. They choose five logic levels to represent various voltage ranges in logic gates:

1. (0): Hard zero

2. (0*): Soft zero

3. (I): Indeterminate, near the logic threshold

4. (1*): Soft one

5. (1): Hard one

Logic level 1 represents a hard one and logic level 1* represents a soft one which is recognized as logic one by a fanout logic gate but can not drive an NMOS pass transistor to fully on state. Logic level I represents an indeterminate level near the logic threshold of an inverter. Such a level may be interpreted as 0, or 1, or I (i.e., the output of the gate is also indeterminate) by the following gate. Logic level hard 0 is always interpreted as logic 0. Finally, logic 0* corresponds to a soft zero, i.e., if it is applied to a dynamic latch, (NMOS pass gate and output capacitance), it can discharge any stored charge on the drain in a time comparable to the propagation delay of the transistor switch, provided source is grounded [2,3].

3.2.2.1 NMOS Logic Gates

Fig. 3.3(a) depicts a three input NMOS NAND gate with various failures locations marked. Fault 7 refers to an open anywhere in the conducting path from output terminal to the ground due to open in the interconnect or due to a missing contact. Fault 1 refers to a short between the gate and the drain of a transistor. Similarly, Faults 2-6 represent shorts between terminals of transistors. Fault 8 represents an open in the path from output to VDD.

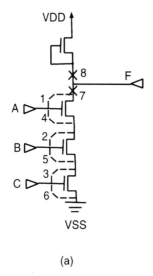

Inputs			Outputs					
A	B	C	F_0	F_1	F_2	F_3	F_4	F_5
1	1	1	0	I	I	I	1	0
0	1	1	1	0	1	1	1	Q^n
1	0	1	1	1	0	1	1	Q^n
1	1	0	1	1	1	0	1	Q^n

O^n means previous logic state is stored
F_0 is the normal fault-free output
Fault classes and faults: $F_1 = \{1\}$; $F_2 = \{2\}$;
$F_3 = \{3\}$; $F_4 = \{4,5,6,7\}$; $F_5 = \{8\}$.

(a)   (b)

Fig. 3.3: NMOS NAND gate (a); and tests for various failures [6].

Floating gate failures (such as a defect on input A, B or C) can be modeled in many ways. For enhancement mode NMOS switching transistors, the presence or absence of trapped charge in the gate (thin) oxide will cause transistor to be either stuck-on or stuck-off. This behavior is equivalent to a SA1 or SA0 fault on the gate terminal. However, in most cases, if the failure is permanent, the stored charge will eventually leak away through the leakage path from the gate terminal to the substrate [3]. Therefore, the substrate and gate voltage will remain the same and the enhancement mode transistor will always be off. As a result such defects in NMOS NAND gate can be represented as fault 7 of Fig. 3.3(a). On the other hand, for depletion mode NMOS load transistor, the same failure will appear as low supply of charging current resulting in a long charging time. The floating gate will not give any logical error and will cause timing errors.

Fig. 3.3(b) illustrates fault-free (F_0) and fault class responses (F_1 - F_5) of the input test vectors in a tabular form. If two defects produce the same faulty response, they are put together in a fault class. Q^n represents the high impedance state where previous logic state is stored. If all inputs are driven by logic gates and are logic high, then the output of the NAND will be 1* for defects 1,2, and 3. However, if inputs are driven through pass transistors, then the output of NAND will be I which may be interpreted as a 0 or a 1 by the following logic gate. Therefore, this is not a reliable test for the above mentioned defects, nevertheless, it is a good test for defects 4,5,6, and 7. It is visible from the table that the stuck-at test set (111, 011, 101, 110) detects all above mentioned fault classes. Similarly, for a three input NMOS NOR gate, a particular sequence of stuck-at test set (100, 010, 001, 000) detects all fault classes similar to those in the illustrated NAND gate.

3.2.2.2 CMOS Logic Gates

Defects in CMOS circuits have similar fault behavior compared to NMOS circuits except for a class of open defects. This class of open defect in CMOS logic gates causes logic gate to have memory like behavior for certain input conditions (Chapter 2). Fig. 3.4(a) illustrates a three input NAND gate with possible failure locations and 3.4(b) depicts the fault detection table. Readers should notice that although in CMOS implementation, compared to NMOS implementation, the number of transistors has increased from four to six, however, the number of fault classes has more than doubled. In general, it is more difficult to test a CMOS logic gate compared to a NMOS logic gate for likely defects.

As it appears the short defects are detected by the stuck-at test set, however, open defects require a particular sequence for detection. Therefore, stuck-at test set is not enough and a test vector sequence of 111, 011, 111, 101, 111, 110 is needed to detect all modeled defects. In general, test vectors for an arbitrary CMOS logic gate may be generated in a manner explained below. First, all primitive sub-networks in the pull-down network are identified. Defects in pull-down sub-networks are detected by the following test sequence [3]:

1. Apply test vector producing logic high at the gates of N channel transistors in the sensitized N channel sub-network. This test detects open defects at

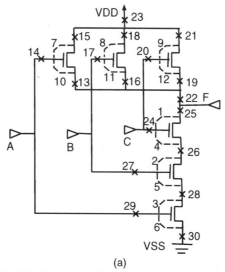

(a)

Inputs			Outputs												
A	B	C	F_0	F_1	F_2	F_3	F_4	F_5	F_6	F_7	F_8	F_9	F_{10}	F_{11}	F_{12}
1	1	1	0	I	I	I	1	0	0	0	0	0	0	Q^n	0
0	1	1	0	1	1	I	1	0	1	1	Q^n	1	1	1	Q^n
1	1	1	0	I	I	I	1	0	0	0	0	0	0	Q^n	0
1	0	1	1	1	I	1	1	1	0	1	1	Q^n	1	1	Q^n
1	1	1	0	I	I	I	1	0	0	0	0	0	0	Q^n	0
1	1	0	1	I	1	1	1	1	1	0	1	1	Q^n	1	Q^n

Fault classes and faults: $F_1 = \{1,12\}$ $F_2 = \{2,11\}$ $F_3 = \{3,10\}$; $F_4 = \{4,5,6\}$; $F_5 = \{7\}$; $F_6 = \{8\}$; $F_7 = \{9\}$; $F_8 = \{13,14,15\}$; $F_9 = \{16,17,18\}$; $F_{10} = \{19,20,21\}$; $F_{11} = \{24,25,26,27,28,29,30\}$; $F_{12} = \{22,23\}$.

(b)

Fig. 3.4: CMOS NAND gate (a); and tests for various failures (b) [6].

all three terminals of the transistor in question, and source/drain short in the transistor. This test will also detect shorts between gate and drain of corresponding P channel transistor.

2. Next, test vector producing logic low at the gate of N channel transistors in N channel sub-network and logic high on the gates of other N channel transistors is applied. In this way the N channel transistor whose gate is set at

logic low is tested for shorts between gate and drain. This test also detects the source/drain short and open defects of all three terminals of the corresponding P channel transistor in the pullup network. This procedure is repeated for every transistor in the N channel sub-network of the logic gate.

The above mentioned test procedure is a typical example of transistor level test generation. Here, it is pertinent to mention that defects considered in this analysis are very simplistic and zero and infinite impedances are assumed for shorts and open defects, respectively. We shall see in subsequent studies that this assumption in not entirely correct and actual defect detection is more difficult.

3.2.3 Study by Maly, Ferguson and Shen

Maly, Ferguson and Shen [16] analyzed the impact of physical defects over NMOS and CMOS cells and developed a systematic methodology [22] for such an analysis. The main difference between their study and the studies reported earlier was the determination of physical origin of the defect and subsequent modeling at appropriate level of abstraction for the fault simulation purpose. A MOS technology consists of a set of layers to be processed through masks. Each mask discriminates between areas to be processed and not to be processed on a given layer. The processed area is assigned the value one and unprocessed area is assigned zero. In this way, it is possible to make a set of Karnaugh maps for the active area, poly, or transistor, etc. In other words, an ideal signature for a given process is determined.

A major cause of defects is the improper processing of layers, or improper interaction among different layers. There could be several causes resulting in improper processing. However, what is important is to be able to put improper processing step onto Karnaugh maps and establish defect and fault relationship. In this way realistic faults could be obtained. Once a fault and its impact on the circuit is known a test could be developed for it. This process is explained with the help of Fig. 3.5. The figure illustrates two layers, A and B. The area to be processed on these layers is marked with 1 and rest of the layer

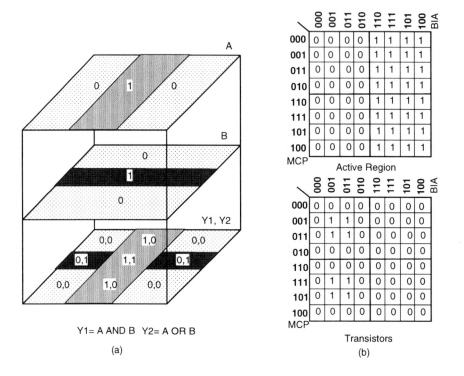

Fig. 3.5: Example of topological operation (a); examples of Karnaugh maps (b).

with 0, respectively. If a logic AND operation is performed over layers A and B, it results in an area that is common in both layers. This layer is depicted as Y1 (coded as 1,1). Similarly, if a logic OR function is performed over A and B, it results in area depicted on layer Y2 (coded as 1,0 OR 0,1). In a similar fashion more complex Boolean relationships can be summarized amongst different layers using Karnaugh maps. Fig. 3.5(b) illustrates Karnaugh maps for active region and transistors, respectively. The axes of Karnaugh map represent different masks. For example, M, C, and P represent metal, CVD SiO2, and polysilicon, respectively. Similarly, B, I, and A represent thin (gate) oxide, transistor implant, and thick (field) oxide, respectively.

The electrical equivalence between different combinations representing the shapes of layers (or processing steps) can also be determined using Karnaugh maps in a similar manner. In this fashion, defects at the processing level of abstractions are translated to the device level of abstraction using a table look-up approach. Such a table can be developed along with the Boolean functions for equivalence classes. The impact of a defect at device level can be determined using circuit level simulator. Although this is a time consuming process, however, in many cases a higher level of simulator may be used without a significant loss of accuracy.

An NMOS full adder was analyzed through this procedure. Both, presence of extra material and absence of material, was considered on metal, diffusion, and polysilicon layers. Therefore, the fault model included six types of spot defects. Based on this fault model, a total of 734 spot defects in the NMOS full adder were generated. These defects were analyzed. Results of this analysis at the circuit level are shown in Table 3.2(a). This table includes only the defects that can be translated into logic level faults. The first category represents single SA faults. Such faults are typically shorts between a circuit node and VDD (VSS). Similarly, category four represents bridging faults which short two (or more) internal circuit nodes. Categories 2 and 3, Source/drain shorts and opens, represent transistor level faults and are self explanatory. Mixed faults are multiple faults which have more than one previously described categories. For example, a missing diffusion may cause a particular transistor gate to be at SA0 and other transistor to be stuck-open. However, such faults are less likely and only one fault is reported. Finally, all defects which create an open circuit on VSS or VDD are classified as power faults.

Table 3.2(b) illustrates the results of logic level fault simulation over total number of logic level significant faults (93). The single SA0/1 row gives the number of defects of each type that can be modeled as a single line SA0 or SA1 (*Readers should not have difficulty in identifying these faults as the tran-*

Table 3.2: Results of the defect simulation at the circuit level (a); and at the logic level (b).

	Extra Metal	Extra Diff.	Extra Poly.	Miss. Metal	Miss. Diff.	Miss. Poly.
simulated	110	64	130	138	145	147
Single SA faults	7	1	2	0	7	1
Source/drain shorts	0	2	0	0	0	7
Source/drain opens	0	0	0	0	8	0
Bridging faults	4	0	12	0	0	0
Floating lines & gates	0	0	1	16	1	17
Mixed faults	0	0	0	0	1	1
Power faults	0	0	0	5	0	0
Total with logical significance: 93	11	3	15	21	17	26

(a)

	Extra Metal	Extra Diff.	Extra Poly.	Miss. Metal	Miss. Diff.	Miss. Poly.
Total: 93	11	3	15	21	17	26
Can be modeled as SA0/SA1: 65	7	3	2	15	13	25
Can be modeled as Single cell I/O SA0/SA1: 31	5	1	2	12	8	3

(b)

sistor level SA faults). While performing this analysis two basic assumptions are made:

1. The circuit level schematic is known. A logic level schematic with best correspondence to the circuit level schematic is evolved such that circuit level faults can be mapped at logic level.

2. Floating lines and gates are assumed to remain at a fixed value (0 or 1). This allows a majority, but not all, of the defects resulting in floating lines and floating gates to be modeled as single SAF.

The second row illustrates the number of faults that can be modeled as a single input or output of the cell SAF which is same as the traditional logic level SAF model. This fault model is able to model approximately 35% of the analyzed spot defects. Furthermore, out of 80 possible SAF in the logic schematic of the adder only 37 occurred in this simulation experiment. However, it should be remembered that the defect densities of the six types may vary significantly from process to process. Furthermore, These types do not represent the complete defect spectrum.

3.2.4 Gate Oxide Shorts: Study by Hawkins and Soden

Gate of a MOS transistor is isolated from its channel by a thin layer of an insulating oxide (SiO_2). For a typical 0.5 micron technology, the gate oxide thickness is about 100 A^0 and each successive scaled generation requires further reduction in this thickness. Growing of thin oxide is a very critical and sensitive process step of VLSI manufacturing. For high performance ICs, thin oxide quality and reliability is a major concern [20].

A gate oxide short is an electrical connection through the thin oxide between the gate and any of the other three ports of a MOS transistor (Fig. 3.6). Typically, the resistance of such a short is few kilo-ohms, therefore, such shorts are not detected by structural and functional voltage testing. However, such defects give rise to elevated I_{DDQ} when excited logically. In a study conducted in Sandia National laboratories gate oxide shorts were found to be the major reliability concern. In a high voltage screen experiment conducted over more than 5,000 1k CMOS static RAMs, 687 devices failed due to Time Dependent Dielectric Breakdown (TDDB) of the gate oxide and 254 (37%) of these devices failed the I_{DDQ} test but passed the functional test. The other 433 devices failed the I_{DDQ} as well as functional tests [11,23].

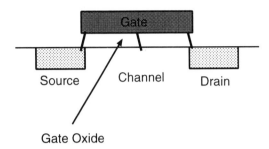

Fig. 3.6: Structure of a MOS transistor and possible gate oxide short sites.

A gate oxide short may result from many reasons. Shorts between the gate and source or drain are often caused by electrostatic discharge or electrical overstress. Furthermore, gate oxide shorts may also occur during the fabrication process. In addition, these shorts can occur later when defects in the oxide results in oxide break-down because of electrical field and thermal stress. The delayed occurrence of the gate oxide short is often referred to as TDDB. There are two dominant causes of TDDB: (i) **Defect or contaminant based failures-** Silicon surface impurities or imperfections cause a local thinning of thin oxide. A local thinning of gate oxide thickness results in higher electric field across the spot that further damages the thin spot. This positive feedback process continues with time till a complete break-down occurs. (ii) **Hot electron based failures-** Hot electrons cause damage to the gate oxide resulting in trapped charge in the gate oxide which attracts more hot electrons causing further damage. This process also continues till it results in a complete break-down of the oxide [12,20,23].

Hawkins and Soden used electrostatic discharge (ESD) and laser techniques to create gate oxide shorts to study the properties of such defects. ESD creates higher electric field near the edge of the gate and typically produces gate oxide shorts between gate and source/drain of a transistor. For an n-doped poly and n+ diffusion such defects cause ohmic connections with values ranging from 800 ohms to 4k ohms. Contrary to ESD, the laser technique could create

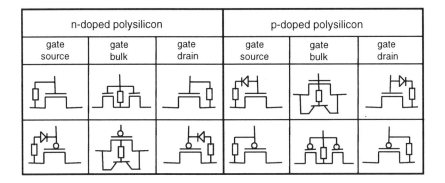

Fig. 3.7: A generalised gate oxide defect model [21].

gate oxide defects in any region. Hence, it was utilized to study defects be-tween n-doped poly and p-well. Such defects formed a *pn*-junction between the terminals. Similarly, gate oxide shorts between n-doped poly and source/drain in PMOS transistors formed a *pn*-junction and gate oxide shorts between poly and n-type substrate formed a resistor. Unless a gate oxide defect results in a complete oxide breakdown, a gate oxide defect typically does not prevent transistor from performing its logical operation. However, its performance (rise or fall time) is degraded substantially. Hawkins and Soden reported an average of 29% reduction in transistor transconductance by a gate oxide de-fect. As reported earlier, such defects result in abnormally high I_{DDQ} and, therefore, are easily detected by I_{DDQ} measurements.

A substantial research effort has been devoted to the modeling of gate oxide defects [21,25]. In general, such defects are either ohmic or rectifying depend-ing upon the polarity (p-type or n-type) of the two shorted nodes. Fig. 3.7 il-lustrates a generalized gate oxide defect model for MOS transistors.

3.3 IFA Experiments on Standard Cells

In another study, Ferguson and Shen [9] extracted and simulated CMOS faults using the IFA technique. A CAD tool, FXT, was developed with a capability of automatic defect insertion/extraction for a reasonably large layout. This tool was used to analyze five circuits from a commercial CMOS standard cell library. The five circuits were, (i) a single bit half adder cell, (ii) a single bit full adder cell, (iii) a counter, (iv) another counter, and (v) a 4x4 multiplier. They sprinkled more than ten million defects in two counters which caused approximately 500,000 faults. Similarly, over 20 million defects were sprinkled in the multiplier which caused approximately 1 million faults. Approximately 1/20th of the defects caused faults that conform to the fact that most of the defects are too small to cause a fault. The majority of extracted faults could be modeled as bridging faults, break faults or a combination of these two. For example in the 4x4 multiplier, the bridging and break faults amount to 48% and 42%, respectively. Almost all remaining faults were transistor SON faults which can also be represented as bridging faults between source and drain of transistors. Similarly, a transistor SOP fault is equivalent to a break fault. Therefore, almost all faults could be represented as bridging or break faults or a combination of the two. The only two categories that were not equivalent to above mentioned categories were new transistors and exceptions. A new transistor in the layout is created by a lithographic spot on poly or diffusion mask. Less than 0.7% of faults fall into this category.

The SAF model performed rather poorly in modeling the extracted faults. In the case of the 4x4 multiplier, only 44% of the bridging faults could be modeled by SAFs. For non-bridging faults only 50% of the faults could be modeled as SAFs. Hence, for the multiplier, less than 50% of all extracted faults could be modeled as SAFs. A similar comparison is carried out with graph-theoretic (transistor-level) fault models. It is estimated that only 57% of the extracted faults could be modeled as graph-theoretic fault model. Though this is higher than those modeled by the SAF model, the majority of non-SAF extracted faults could not be modeled. Two reasons were attributed to this. Firstly, many non-SAF faults bridge input nodes together and are not modeled with the graph-theoretic approach either. Secondly, approximately 70% of the transistors in the analyzed circuits were pass transistors or components of inverters. Pass transistors are not modeled in graph-theoretic fault modeling and

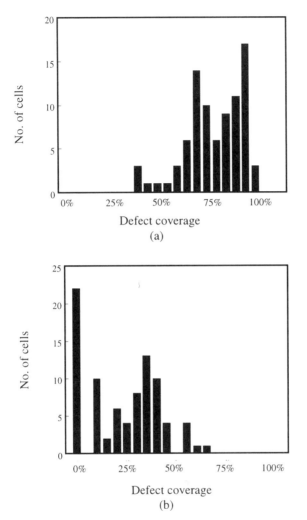

Fig. 3.8: Voltage coverage of cell library with low resistive bridging defect
(a); and voltage coverage with 2kilo-ohm resistive bridging defect (b) [19].

SAF model could model most transistor faults occurring in inverters that
caused change in the logical function. Only 1% of the extracted faults could
be represented by the transistor SOP fault model.

Further analysis was carried out to find out how well SA and exhaustive test
sets detect the extracted faults. In order to reduce the simulation time only extracted bridging faults were simulated in counters. The SA test set could de-

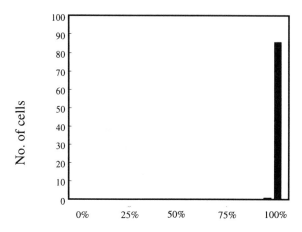

Fig. 3.9: I_{DDQ} bridging fault coverage of a library [19].

tect between 73% to 89% of the circuit's bridging faults. Even under the unrealistic assumption that all non-bridging faults are detected by SA test set, the 100% SA test set could detect between 87% and 95% of extracted faults. The fault coverage of extracted bridging faults by the exhaustive test set was relatively high. The exhaustive test set detected between 89% to 99.9% of the extracted bridging faults. As a test solution for better fault coverage quiescent current monitoring (I_{DDQ}) was suggested. It was implied that I_{DDQ} will provide the best test set for bridging fault detection.

Peters and Oostdijk [19] analyzed Philips standard cell library using six types of leakage faults. Their analysis was restricted to bridging defects in standard cells. They did not consider open defects, assuming that the probability of occurrence of open defects is relatively small. The bridging fault coverage of the library by voltage and I_{DDQ} tests is illustrated in Fig. 3.8 and Fig. 3.9, respectively. Fig. 3.8(a) depicts that for low resistive defects (typically less than 10 ohms) no cell has a 100% coverage for the modeled faults by voltage testing. Most cells have voltage coverage between 60% - 90% for the modeled faults. Fig. 3.8(b) illustrates the voltage fault coverage when the bridging fault resistance is 2k ohms. As expected, the voltage fault coverage declines rapidly as the defect resistance is increased. A large number of cells (22) was reported with zero fault coverage.

The I_{DDQ} coverage of the leakage fault model is 100% for all combinational cells for a threshold current of 10 μA. Sequential cells do not have 100% I_{DDQ} fault coverage. It is a known fact that a class of bridging defects in latches and flip-flops do not give rise to elevated I_{DDQ}. *In Chapter 4, we shall discuss the defects in sequential circuits and suggest measures to improve the defect coverage.*

An analysis of the results stresses that I_{DDQ} testing is a necessity for reliability since it detects bridging and transistor short defects. Even assuming zero ohm defect resistance, the defect coverage of voltage vectors can not be expected to be higher than about 75%. An analysis of the detailed results has shown that especially defects on serial transistors and defects between inputs are hard to detect using a voltage test. These defects are easily detectable using I_{DDQ} testing. Also, many GOS defects can not be detected using voltage vectors, which could be a possible explanation for early life failures.

For minimizing the overlap between detection by voltage or I_{DDQ} vectors, the use of a critical resistance is very important. Defects with resistance below the critical resistance will be detected by voltage testing. However, detection of defects above the critical resistance is not assured. Introduction of a threshold resistance value for voltage testing will assure the detection of defects up to a particular resistance value will be detected. In other words, based on defect resistance statistics for a given process, one can quantify how I_{DDQ} testing will improve the defect detection.

3.4 I_{DDQ} Versus Voltage Testing

Logic testing, structural as well functional, has been the cornerstone for IC testing. In the last decade, I_{DDQ} testing has been increasingly used as a quality improving supplement to the logic testing. A number of studies were reported

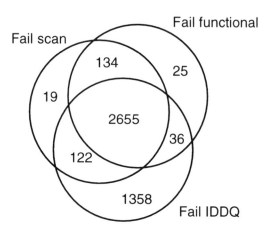

Fig. 3.10: Distribution of failing die in each test class reported by Maxwell et al. [17].

on the relative effectiveness of both of these test techniques [1,17,18,24]. Some aspects are discussed below briefly to give an idea to the reader.

Perry [18] reported a three year study of CMOS ASICs. I_{DDQ} testing was implemented to reduce the early life failures. A set of 13 ASICs were analyzed in this study and typical SA fault coverage of devices was more than 99%. It was demonstrated in the study that with the implementation of I_{DDQ} the rejection rate was reduced at least by a factor of four. Similarly, Maxwell et al. [17] conducted a study of three wafer lots containing 26,415 die (excluding parts which failed initial continuity and parametric tests). The distribution of failing die in each of the test class is illustrated in Fig. 3.10. Most defective chips (2655) were identified by all tests. A large number of defective chips (1358) was detected only by I_{DDQ} test and 25 and 19 failures were only detected by functional and scan tests, respectively. Table 3.3 shows the reject rates which would result for various combinations of the tests. If no testing is done, the reject rate would be 16.5% while if just functional and I_{DDQ} tests were performed, the rate would be 0.09% (900 PPM). If only the I_{DDQ} test was to be performed the reject rate would be 0.80%.

Table 3.3: Reject rate for various tests reported by Maxwell et al. [16].

Reject Rates (%)	Scan and Functional Tests			
	Neither	Noscan/ Func	Scan/ Nofunc	Both
Without IDDQ	16.46	6.36	6.04	5.80
With IDDQ	0.80	0.09	0.11	0.00

Aitken [1] investigated the potential of I_{DDQ} testing in defect diagnosis. He showed that using both inter and intra-gate shorts as fault models and measuring the current under different steady state input conditions, it is possible to diagnose a defect location and/or determine its cause. This hypothesis was applied to ASICs and out of 151 parts in the sample, diagnosis was obtained for 135. In many cases, the predicted defects were confined to a single standard cell. The transistor short model (transistor leakage fault model) could diagnose 120 of the diagnosed defects. The input SA fault model was second with 90, however, all of them were also detected by transistor short fault model. Inter-gate bridge fault model could diagnose only 30 of these out of which 15 were only detected by this fault model. A total of 16 failed devices could not be modeled by any of the above mentioned fault models. The success of the transistor short model may be due to the fact that the I_{DDQ} test is the only one that specifically targets those faults. Furthermore, these results are biased towards I_{DDQ} test since failing parts had passed all tests except I_{DDQ} tests.

Subtle defects such as resistive transistor faults usually change the transistor transfer characteristics. Such changes in the transfer characteristics may cause an increased transistor delay or an increased I_{DDQ} current. Vierhaus et al. [26] carried out extensive simulations to quantify the impact of resistive stuck-on, stuck-open, and bridging faults on delays and I_{DDQ} for typical CMOS logic gates. The results of their analysis were more or less predictable for I_{DDQ} testing. All resistive transistor faults gave rise to state dependent elevated I_{DDQ}. However, their analysis also highlighted certain faults that reduced the logic

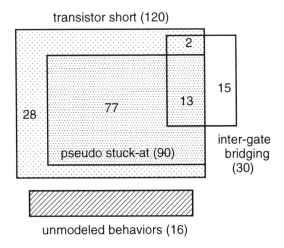

Fig. 3.11: Overlap of faulty behaviors observed by Aitken [1].

delay. It may be recalled that usually bridging faults degrades the switching characteristics and increase the transistor/logic delay. This phenomena can be further explained with the help of Fig. 3.12. This figure illustrates a two input NAND gate with a bridging fault between an input (In1) and output (Out). Such a fault may be caused either by low a resistive polysilicon bridge or a high resistive oxide defect through the thin oxide of the particular transistor (gate-drain gate oxide defect). Typically, a low resistive bridge causes a feedback condition resulting in a functional fault. A strong input driver, via the bridge, directly drives the faulty gate's output node. In the case of a weak driver, the output of faulty gate switches at an intermediate voltage (usually close to VDD/2), depending upon the input. As the fault resistance is increased (>10k ohms), the fault causes positive delay and/or SAF. However, as the fault resistance is increased above 10 kilo-ohms, a positive feedback effect is created.

The resistive transistor stuck-on and stuck-open faults also behave somewhat similarly. A low resistive stuck-on fault in a PMOS transistor of a NAND gate will most likely cause a SAF behavior. As the resistance of the faulty transistor rises, it causes delay fault in the NAND gate. A stuck-on fault in an NMOS transistor usually does not result in a SAF on the output of the NAND gate.

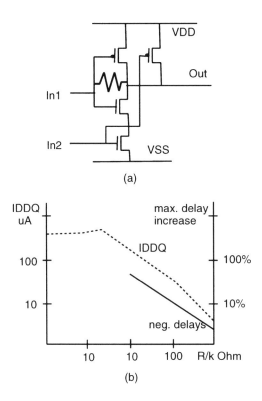

Fig. 3.12: A bridging fault improving logic delay of a two input NAND gate [24].

The fault supported transitions result in smaller delays. All these faulty conditions are detected by I_{DDQ} measurements. The faulty behavior of CMOS gates under non-ideal stuck-open conditions (transistor source/drain open) with approximate resistance value of 50k ohms results in gross delay fault. Such defects are not detected by I_{DDQ} measurements.

3.5 Defect Classes and their Testing

Fault models represent defective behaviors with limited accuracy. We have seen in previous sections and the previous chapter that there is no single test method suitable for testing all possible defect types. Furthermore, if the test objective is to guarantee a low escape rate (low PPM), then the test should be

Table 3.4: CMOS IC Defect classes [13].

Defect Class	Description	Test Method	100% Detect
Bridge Type-1	Trans. node, interlogic gate, logic gate to power bus, power bus-to-bus	I_{DDQ} Boolean	Yes No
Bridge Type-2	Layout identified bridges	I_{DDQ} & Boolean	Yes
Bridge Type-3	Sequential intranodel	I_{DDQ} & Boolean	Yes
Open Type-1	Transistor-on	I_{DDQ} Boolean	Yes No
Open Type-2	Transistor pair-on	I_{DDQ} Boolean	Yes Yes
Open Type-3	Transistor pair-on/off	Boolean I_{DDQ}	Yes No
Open Type-4	Sequential	Boolean & I_{DDQ}	Yes
Open Type-5	Transistor-off (Memory)	I_{DDQ} Boolean	No No
Open Type-6	Delay	I_{DDQ} Boolean	No No
Parametric Delay	R_{via}, V_T, $\Delta(W/L)$	I_{DDQ} Boolean	No No

focused on defects rather than on faults. Hence, it is of vital importance to categorize types of defects in CMOS circuits and outline their detection strategies. Keeping this objective in mind Hawkins et al. [13] identified defect classes. Defects can be segregated into different types of bridges and opens. Table 3.4 illustrates these types.

It is apparent from the table that several defect classes are 100% detectable considering that appropriate test vectors are either available or can be generated. I_{DDQ} is the most effective test for bridging defect classes with Boolean testing also achieving 100% coverage for type 2 and 3 bridges assuming that the defect resistance is lower than 1k ohms. If the defect resistance increases beyond 1k ohms, the Boolean coverage of defects is reduced.

An open defect causing a transistor stuck-on behavior (Open Type-1) is 100% detectable using pseudo stuck at fault (PSAF) patterns. Boolean detectability is difficult since stuck-on behavior causes degraded voltage levels at logic output. Open Type-2 defects are 100% detectable by I_{DDQ} tests as well as by Boolean testing with PSAF patterns. If an open defect causes a transistor-pair on/off then Boolean test will detect it and I_{DDQ} will not detect it. Open defects in sequential circuits will be detected by Boolean as well as I_{DDQ} tests. In order to detect Open Type-5 a 2-pattern sequence is needed and simple SAF testing or I_{DDQ} testing will not be effective. Marginal opens (Open Type-6) are hard to detect by Boolean or I_{DDQ} testing. Similarly, parametric faults are hard to detect by both methods. In order to detect such faults delay fault testing or at speed testing seems to be promising alternatives.

3.6 Conclusion

The popularity of stuck-at faults has led to several studies to determine whether SAF model represents manufacturing defects. Results of various studies are mixed. Galiay et al. found that only a subset of defects resulted in logical faults. The study of Banerjee and Abraham concluded that the transistor level fault models represent transistor defects with fair amount of accuracy. However, in their study they assumed zero and infinite impedances for shorts and opens, respectively. With more realistic defect impedances their fault coverage would have been lower. A MOS technology is a collection of a set of layers to be processed through masks and each mask discriminates between areas to be processed or not processed on a given layer. A defect on a mask or a dust particle on a wafer may result in the improper processing of any layer. By performing logical operations on the masks of different layers defects are abstracted at the device level. In this way realistic faults are generated. Maly et. al. proposed a methodology for the same and carried out an analysis for NMOS and CMOS circuits. The conventional SAF model could represent only 35% of all defects that could occur and out of 80 possible SAFs in a given circuit only 37 actually occurred. Similarly, studies conducted on standard cells by Ferguson and Shen, and Peters and Oostdijk, respectively, demonstrated that SAF model is a poor abstraction of realistic defects. Their studies concluded I_{DDQ} is far more effective in defect detection.

The realization of gate oxide is one of the most critical process steps. Each successive scaled generation has still thinner gate oxide and the quality of gate oxide often determines the product quality and reliability in the field. Hawkins and Soden conducted an experiment on SRAMs where 37% of faulty devices (poor gate oxide) passed all functional tests but failed I_{DDQ} tests. Several subsequent studies were conducted on I_{DDQ} and logic testing. The results of these studies were overwhelmingly in favor of I_{DDQ} testing.

References

1. R. C. Aitkens, "A Comaprison of Defect Models for Fault Location with I_{DDQ} Measurements," *Proceedings of International Test Conference,* 1992, pp. 778-787.

2. P. Banerjee and J.A. Abraham, "Fault Characterization of VLSI Circuits," *Proceedings of IEEE International Conference on Circuits and Computers*, September 1982, pp. 564-568.

3. P. Banerjee and J.A. Abraham, "Characterization and Testing of Physical Failures in MOS Logic Circuits," *IEEE Design & Test of Computers*, vol. 1, pp. 76-86, August 1984.

4. C.C. Beh, K.H. Arya, C.E. Radke and K.E. Torku, "Do Stuck Fault Models Reflect Manufacturing Defects?" *Proceedings of International Test Conference*, 1982, pp. 35-42.

5. R. Bryant, "A Survey of Switch-Level Algorithms," *IEEE Design & Test of Computers,* vol. 4, pp. 26-40, August 1987.

6. R. Chandramouli, "On Testing Stuck-Open Faults," *Proceedings of 13th Annual International Symposium on Fault Tolerant Computing Systems*, June 1983, pp. 258-265.

7. K.W. Chiang and Z.G. Vranesic, "Test Generation For MOS Complex Gate Networks," *Proceedings of 12th Annual International Symposium on Fault Tolerant Computing Systems*, June 1982, pp. 149-157.

8. Y.M. El-Ziq and R.J. Cloutier, "Functional Level Test Generation for Stuck-Open Faults in CMOS VLSI," *Proceedings International Test Conference,* 1981, pp. 536-546.

9. F.J. Ferguson and J.P. Shen, "Extraction and Simulation of Realistic CMOS Faults using Inductive Fault Analysis," *Proceedings of International Test Conference,* 1988, pp. 475-484.

10. J. Galiay, Y. Crouzet and M. Vergniault, "Physical Versus Logical Fault Models in MOS LSI Circuits: Impact on Their Testability," *IEEE Transaction on Computers,* vol. C-29, no. 6, pp. 527-531, June 1980.

11. C.F. Hawkins and J.M. Soden, "Electrical Characteristics and Testing Considerations for Gate Oxide Shorts in CMOS ICs," *Proceedings of International Test Conference,* 1985, pp. 544-555.

12. C.F. Hawkins and J.M. Soden, "Reliability and Electrical Properties of Gate Oxide Shorts in CMOS ICs," *Proceedings of International Test Conference,* 1986, pp. 443-451.

13. C. F. Hawkins, J. M. Soden, A. Righter, and J. Ferguson, "Defect Classes - An Overdue Paradigm for CMOS IC Testing," *Proceedings of International Test Conference,* 1994, pp. 413-425.

14. S.K. Jain and V.D. Agrawal, "Modeling and Test Generation Algorithm for MOS Circuits," *IEEE Transactions on Computers,* vol. 34, no. 5, pp. 426-43, May 1985.

15. N.K. Jha and S. Kundu, *Testing and Reliable Design of CMOS Circuits,* Boston: Kluwer Academic Publishers, 1990.

16. W. Maly, F.J. Ferguson and J.P. Shen, "Systematic Characterization of Physical Defects for Fault Analysis of MOS IC Cells," *Proceedings of International Test Conference,* 1984, 390-399.

17. P. C. Maxwell, R. C. Aitken, V. Johansen and I. Chiang, "The Effectiveness of I_{DDQ}, Functional and Scan Tests: How Many Fault Coverages Do We Need?," *Proceedings of International Test Conference,* 1992, pp. 168-177.

18. R. Perry, "I_{DDQ} Testing in CMOS Digital ASIC's - Putting It All Together," *Proceedings of IEEE International Test Conference,* 1992, pp. 151-157.

19. F. Peters and S. Oostdijk, "Realistic Defect Coverages of Voltage and Current Tests," *Proc. of IEEE International Workshop on I_{DDQ} Testing*, 1996, pp. 4-8.

20. K. F. Schuegraf and C. Hu, "Reliability of thin SiO_2," *Proceedings of IEE*, vol. 9, pp. 989-1004, September 1994.

21. J. Segura, C. Benito, A Rubio and C.F. Hawkins, "A Detailed Analysis of GOS Defects in MOS Transistors: Testing Implications at Circuit Level," *Proceedings of International Test Conference*, 1995, pp. 544-551.

22. J.P. Shen, W. Maly and F.J. Ferguson, "Inductive Fault Analysis of MOS Integrated Circuits," *IEEE Design & Test of Computers*, vol. 2, pp. 13-26, December 1985.

23. J. M. Soden and C.F. Hawkins, "Test Considerations for Gate Oxide Shorts in CMOS ICs," *IEEE Design & Test of Computers,* vol. 3, pp. 56-64, August 1986.

24. T. Storey, W. Maly, J. Andrews, and M. Miske, "Stuck Fault and Current Testing Comparison Using CMOS Chip Test," *Proceedings of International Test Conference,* 1991, pp. 311-318.

25. M. Syrzycki, "Modeling of Gate Oxide Shorts in MOS Transistors," *IEEE Transactions on Computer Aided Design,* vol. 8, no. 3, pp. 193-202, March 1989.

26. H.T. Vierhaus, W. Meyer, and U. Glaser, "CMOS Bridges and Resistive Faults: I_{DDQ} versus Delay Effects," *Proceedings of International Test Conference*, 1993, pp. 83-91.

27. R.L. Wadsack, "Fault Modeling and Logic Simulation of CMOS and MOS Integrated Circuits," *Bell Systems Technical Journal*, vol. 57, no.5, pp. 1449-1474, May-June 1978.

| CHAPTER 4 | # Testing Defects in Sequential Circuits |

Scan chains are a popular feature in complex ICs. The inherent test complexity of a scan chain is linear with respect to the number of flip-flops it contains. In this chapter, a DFT strategy is outlined such that a scan chain can be tested with very few test vectors by making the chain transparent in the test mode. Such a strategy improves the defect coverage of the I_{DDQ} test for flip-flops. For ICs without scan chains, robust flip-flops are proposed that are inherently I_{DDQ} testable.

4.1 Introduction

In previous chapters, we mentioned how scan path, LSSD and their derivatives became popular because their application could change distributed sequential logic elements into a big unified shift-register for testing purposes [5,8]. As a result, the overall test complexity is reduced [28]. Owing to these techniques, test generation and fault grading for complex digital circuits became a possibility.

At the same time, a number of studies revealed that classical voltage based test methods for digital CMOS ICs are inadequate in ensuring the desired quality levels [6,15]. Increased quality awareness brought in new test techniques like quiescent current measurements (QCM), or I_{DDQ} (I_{SSQ}) as it is popularly known, in the test suite for digital CMOS ICs. I_{DDQ} testing has been shown to be very effective in quality improvement and in reducing test cost and complexity for static digital CMOS ICs. This technique makes use of an important property of CMOS circuits that in steady state current consumption is very small. Therefore, a higher steady state current is an indicator of a probable defect/fault. Furthermore, the quality levels obtained by QCM can not be obtained otherwise by any other test method [7,12,23-25]. However, QCM method is not without its limitations. For example, only a subset of open defects is detected by QCM. Compared to the voltage test method, QCM is considerably slower. Moreover, general purpose ATE is not tailored for fast QCM. Recently, there have been some efforts to put a dedicated I_{DDQ} monitor on an ATE to speed up the measurement. A quality test action group (QTAG) is formed to address I_{DDQ} and ATE related aspects [1].

Unlike combinational circuits, the controllability condition in sequential circuits (e.g., flip-flops) does not ensure that a bridging fault (short defect) is detected by I_{DDQ} testing. Many low resistance shorts in sequential circuits do not cause an elevated I_{DDQ}. The voltage detection of such defects depends on transistor level parameters (e.g., width and length) of the affected transistors and resistance of the defect. Therefore, in this chapter, we concentrate on defects in sequential logic gates and propose some testability solutions. In the following section, the problems with bridging fault detection in sequential circuits are illustrated. Section 4.3 gives a local solution and demonstrates that most bridging faults in a flip-flop can be detected by I_{DDQ} as well as by voltage tests if in the test mode the flip-flop is made transparent. In Section 4.4 the concept of the flip-flop transparency is extended to scan chains to reduce their test complexity and to effectively test defects. The test complexity of the transparent scan chain is independent of the number of flip-flops. The implementation cost of a chip level solution, Transparent scan (TS), is much lower compared to that of the local solution implementation. Moreover, we shall demonstrate that in the TS mode, most of the spot defects in the scan chain, can be efficiently tested by six test vectors. In many situations, constraints like area, I/O pins, performance, packaging, etc. prevent the use of scan chains in digital ICs. For such cases, inherently I_{DDQ} and voltage testable flip-flop configurations are desirable. In Section 4.5 such configurations are described and their performance is compared with typical CMOS flip-flop configurations. Finally, in Section 4.6 conclusions are given.

4.2 Undetected Defects

Lee and Breuer [11] highlighted a certain class of bridging faults in sequential circuits that are not detected by QCM. Rodriguez et al. [16], carried out inductive fault analysis of a scan flip-flop to find out the relative effectiveness of QCM and voltage test methods for realistic bridging faults detection. Conclusions of their analysis are: (a) For zero resistance bridging faults, 8% can not be detected by QCM. However, these can be detected by output voltage measurements. (b) For bridging faults with resistance above 2k ohms, QCM detected all defects but only some were voltage detectable.

The above mentioned analysis demonstrates that all bridging faults in the analyzed flip-flop are detectable either by voltage or I_{DDQ} test method. Nevertheless, it should be stressed that the detection of such bridging faults by I_{DDQ} or voltage test method strongly depends on the circuit level parameters (e.g., W and L) of transistors in the flip-flop and the resistance of the defect. In a paper, Metra et al. [14] showed that for CMOS flip-flops implemented with NAND gates and/or with pass gates, neither I_{DDQ} testing, nor voltage testing, nor the combination of the two achieved the complete bridging fault coverage. Their study resulted in two broad conclusions: (a) Irrespective of flip-flop implementation, bridging fault coverage in flip-flops is low, and (b) Circuit level parameters have an influence on bridging fault detection. For voltage detection the logic thresholds of intermediate logic gates driven by the flip-flop and the satisfaction of observability conditions will also play an important role before faults can be detected at the primary outputs of the device under test (DUT).

Fig.4.1 shows a typical flip-flop implementation in CMOS technology. It is a single phase clock, master-slave flip-flop. While Clock is at logic low level,

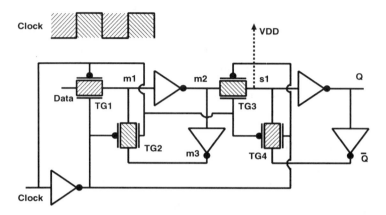

Fig.4.1: A bridging defect in a typical CMOS flip-flop not detected by QCM.

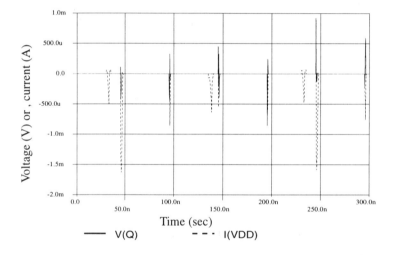

Fig. 4.2: Circuit simulation showing the bridging defect is not detected by QCM.

transmission gates TG1 and TG4 are conducting. At the same time TG2 and TG3 are in the non-conducting state. Hence in this clock state, the master-latch of the flip-flop accepts the new data from the data input (Data) while the slave-latch retains the old data. Following the instance when Clock has a rising transition, the master-latch no longer accepts the input data and transfers the current data to the slave-latch. In this fashion the master-slave operation of the flip-flop is realized.

Extra material defects in metalization layers and gate oxide pin-holes have been identified as the leading defect mechanism in CMOS ICs. These defects are popularly known as shorts or bridging faults. Other defects include open defects, etc. [23,25]. Most of these extra material defects and gate oxide pin-holes, under appropriate steady state input stimuli conditions, give rise to an abnormally high current [24]. Such defects under specific steady-state input stimuli conditions create a DC current path between VDD and VSS and can be detected by QCM. However, the current flowing through the DC path should be higher than the various leakage currents (e.g., transistor leakage) in the IC so as to give the indication of a defect. On the other hand, a defect which can not cause an elevated quiescent current between VDD and VSS in any of the steady-state input stimuli conditions, can not be detected by QCM. A low resistive bridging defect between node s1 and VDD (or VSS) in Fig.4.1 can not cause high quiescent current and therefore is not detected by QCM [14,16].

The problem of bridging fault detection in CMOS flip-flops with QCM is generic in nature [14]. For flip-flops implemented with pass gates, the non-detection of bridging faults is explained as below. In CMOS technology, flip-flops are made economically using switches or TGs. These switches are alternately closed or opened to ensure the master-slave operation of the flip-flop (Fig. 4.1). The reason for non-detection of this bridging fault by QCM is the bidirectional nature of switches. At the rising transition of Clock, TG1 and TG4, which were conducting, stop conducting; and TG2 and TG3, which were not conducting, start to conduct. Now the node m2 starts to drive node s1 which till this moment was driven by node Q through TG4. The input of node m2, itself is going through a transitory phase (since TG1 is turning off and TG2 is turning on), therefore, the node m2 has a limited driving capability. In a defect-free case, positive feedback via a pair of back to back inverters allows the flip-flop to ride through this transitory phase. Now, due to the bridging fault, (Fig. 4.1), the node s1 is constantly driven to VDD (or VSS) level. In the case of a low resistive bridging fault, the voltage driving strength through the defect is much stronger than that of m2 and as a result overrides the master latch.

This operation is similar to the write operation carried on a SRAM cell. Therefore, in steady-state no current flows and the defect is not detected by QCM.
Fig. 4.2 shows the circuit simulation result for the flip-flop with the node s1
shorted to VDD by a low resistive bridging fault. Though it is clear from the
graph that the output voltage shows a stuck-at-0 behavior, no Data and Clock
combinations could create an elevated I_{DDQ}. At Clock transitions a large transition current, similar to the defect-free case, flows for a very short time. This
transition current neither carries the defect information nor it can be measured
by the QCM method. As the resistance of the defect increases, the ability to
drive through the defect reduces. At a certain resistance value, the drive of m2
is comparable to that of the defect. This gives rise to logical conflict which is
QCM detectable [16].

4.3 The Local Solution

As explained in the previous section, a bridging fault is QCM detected if a
logical conflict between faulty nodes can be created and sustained in the
steady-state. In this particular case, the conflict can be kept alive as long as the
master-latch of the flip-flop is not over-written by the drive through the defect.
This can be achieved in one of the following ways [19,20]:

• Initialize the master-latch with appropriate data value keeping Clock low.
  Use an external control signal, Test, that breaks the feedback loop in the
  master-latch. Finally change Clock to logic high so that TG3 starts conducting.

• Maintain the appropriate data value at Data input of the flip-flop. Use an
  external control signal, Test, and Clock, make all TGs conducting at the
  same time.

The first solution can be implemented with an additional TG in the feedback
path of the master-latch, an inverter and a control input, Test. However, the latter solution of making the flip-flop transparent is a better one since all nodes
are driven to either logic high or low. A flip-flop can be made transparent locally if an additional test block is added to it such that in test mode, all four

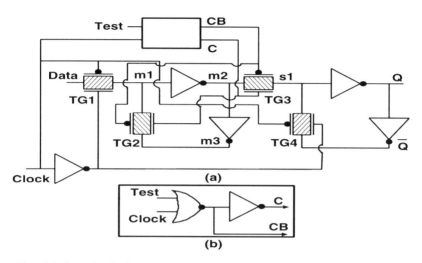

Fig. 4.3: Local solution: (a) concept, and (b) test block implementation.

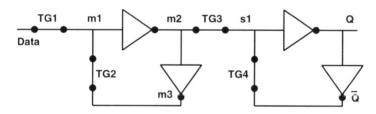

Fig. 4.4: Equivalent circuit of the flip-flop.

TGs of the flip-flop are made to conducting at the same time. This will ensure, that nodes m1, m2 and s1 are always driven by Data input. Fig. 4.3(a) shows the concept, and Fig. 4.4(b) gives a possible implementation of extra test logic required in a flip-flop. Clock and $\overline{\text{Clock}}$ signals to TG2 and TG3 are routed through the test logic. A control signal, Test, is needed to switch between normal and test modes.

The scheme can be implemented with a NOR gate and an inverter. This extra test logic is added to the flip-flop such that when Test is low, the flip-flop behaves exactly as a normal flip-flop. Outputs of the test logic, C and CB, are in fact, Clock and $\overline{\text{Clock}}$. However, in the test mode (Test=1) the C and CB are

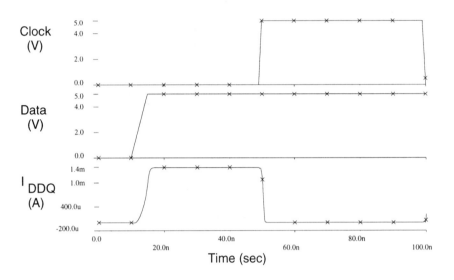

Fig. 4.5: A defect detected by QCM with the application of the
local solution.

held at logic levels high and low, respectively, irrespective of the logic level of
Clock. This ensures that TG2 and TG3 are conducting during the test mode.
Now if Clock is at logic low, TG1 and TG4 are also conducting. Therefore, in
the test mode the flip-flop is transparent. Fig. 4.4 shows the test mode equivalent circuit of the flip-flop. Since in test mode all nodes within the flip-flop are
driven simultaneously, any bridging defect between flip-flop node and VDD
(VSS) node can be detected by QCM as well as by voltage measurement.

Input data value can be setup so as to detect various likely bridging defects by
I_{DDQ} as well as by voltage measurement. Assuming that s1 is shorted to VDD,
Data is also kept high. In this condition there will be a driven logic conflict between nodes s1 and VDD that will give rise to a large quiescent current detected by QCM. Fig. 4.5 shows simulation results. When Clock is low and Data is
high, a large quiescent current can easily be detected by QCM. Similarly, logic low Data will detect the bridging defect between node s1 and VSS. The cost

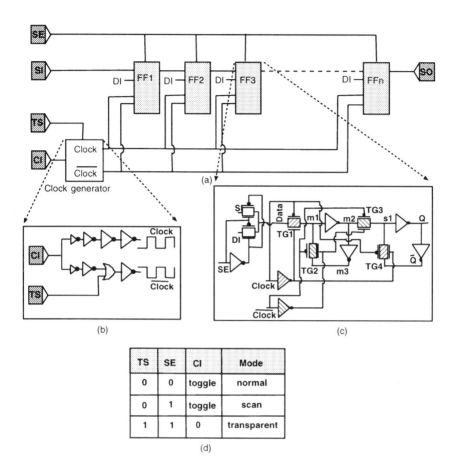

Fig. 4.6: (a) Transparent scan, (b) extra control needed in clock generator, (c) modified scan flip-flop, and (d) modes of flip-flop operations.

of this local implementation is six extra transistors and the test signal routing to each desired flip-flop. A similar solution can be applied to two-phase and level-sensitive flip-flops.

4.4 The Transparent Scan

Although the local solution is attractive in some ways, at chip level a global and a cost effective solution should be applied. For single phase clocked flip-flops, Fig. 4.6 illustrates this concept of scan chain transparency. However, the concept is technology and implementation independent. Fig. 4.6(a) shows a scan chain with a scan input (SI) signal, a scan output (SO) signal, and a scan enable (SE) signal which selects between the scan and the normal mode operations of the scan chain. Fig. 4.6(a) also shows a clock generator with clock input (CI) signal and transparent scan (TS) enable signal. Fig. 4.6(b) and Fig. 4.6(c) show expanded block diagrams of the clock generator and a scan flip-flop, respectively. The Clock generator generates Clock and $\overline{\text{Clock}}$ as outputs. The Clock is controlled by CI and $\overline{\text{Clock}}$ is asynchronously controlled by TS. The flip-flop accepts Clock and $\overline{\text{Clock}}$ as inputs, which are locally inverted again to generate their respective complements. The Clock signal controls TG1 and TG4; and $\overline{\text{Clock}}$ signal controls TG2 and TG3. Fig. 4.6(d) shows various modes of operation in the scan flip-flop. When TS and SE are both 0 and CI is toggling, scan flip-flops accept data from the data inputs (DI). When SE is changed to logic high, all flip-flops function together as a normal scan chain. The scan chain becomes transparent when TS and SE are both at logic high and CI is held at logic low. In this mode the complete scan chain is converted into n pairs of inverters.

For two phase clocked LSSD design style achieving scan chain transparency is considerably simpler. Bottorff et al. [3] proposed this method and named it as the flush test. In LSSD design style, often both phases of the clock can be controlled independently from IC inputs. Therefore, if both phases of the two phase clock are held at active level, the flip-flops, and hence, the scan chain becomes transparent. However, two phase clocking may not be used in many CMOS ICs owing to extra routing and power consumption.

4.4.1 Defect Detection

As mentioned previously, bridging faults are the most probable defects in CMOS ICs. It can be argued that since the complete scan chain is n pairs of inverters, most of the low-resistive bridging faults affecting transistors in the scan chain are detected by output voltage (SO) as well by QCM tests. The detection of high-resistive bridging faults is more difficult. Their detection by

QCM will depend on the resistance of the bridging fault, on-resistance of con-flicting transistors (Rp, Rn) and the QCM threshold. The maximum resistance value of the bridging fault that is detected by QCM is calculated as follows:

$$R_{bridge} + R_p + R_n = \frac{VDD}{QCM_{threshold}} \qquad \text{(EQ 4.1)}$$

Assuming that the threshold of QCM is 10 μA and VDD is 5 V, the total path resistance between VDD and VSS that will give rise to QCM above the threshold can be estimated as 500 kΩ. Typically, the on-resistance of CMOS transistors is in the range of kilo ohms which can be ignored for the calcula-tion of the maximum resistance value of the bridging fault. Therefore, the maximum resistance value of the bridging fault that is detectable by QCM is approximated as:

$$R_{bridge} \approx \frac{VDD}{QCM_{threshold}} \qquad \text{(EQ 4.2)}$$

The determination of maximum detectable value of a bridging fault in a flip-flop is of little practical significance. Nevertheless, the resistance of 500 kΩ that is detectable by QCM is much higher than the reported high-resistive bridging fault [17]. The voltage detection of high-resistive bridging fault will be more difficult. Primarily it will depend upon the bridging fault resistance value with respect to the on-resistance of the conflicting transistor(s) and the condition that subsequent logic gate interprets it as the faulty logic state (Fig. 4.7). This figure is a small subset of Fig. 4.1 with transistor details incorporat-ed. The defect and conducting transistors form a resistive network between VDD and VSS (Fig. 4.7(b)). Due to this resistive network the intended logic low voltage seen by the subsequent inverter can be expressed as follows:

$$V_{low} = VDD \bullet \left(\frac{R_{TG} + R_n}{R_p + R_{bridge} + R_{TG} + R_n} \right) \qquad \text{(EQ 4.3)}$$

If the input voltage, V_{low}, seen by the subsequent inverter is above logic high threshold, the defect will be detected by voltage measurement. The input SI is driven logic high as well as logic low to detect such bridging faults. Since Clock and $\overline{\text{Clock}}$ both are kept at low logic during these two test vectors, (a) a

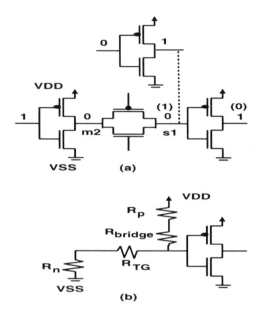

Fig. 4.7: Voltage detection of high resistive bridging faults.

bridging fault between Clock and $\overline{\text{Clock}}$, and (b) a bridging fault between Clock ($\overline{\text{Clock}}$) and VSS will not be detected either by QCM or by voltage measurements. For their detection by QCM, two additional vectors will be required. In the first additional test vector, Clock is kept high while the $\overline{\text{Clock}}$ is kept low, this will ensure the detection of (a). In the second additional test vector, both Clock and $\overline{\text{Clock}}$ are kept high to cover (b). Furthermore, there may be some less likely shorts that can not be detected by this method. For example, a short causing a bridging fault between the source and the drain of a TG will not be detected. Similarly, a short between outputs of two flip-flops will not be detected. However, such defects are unlikely and their probabilities can further be reduced by layout modifications. In the situation when flip-flop outputs form a bus, the bridging faults between the same are likely. In such cases, adjacent bus elements should be driven by Q and $\overline{\text{Q}}$ of different flip-flops so that such bridging faults are detected with QCM. If this is not possible, the voltage test should be devised to explicitly test for such faults.

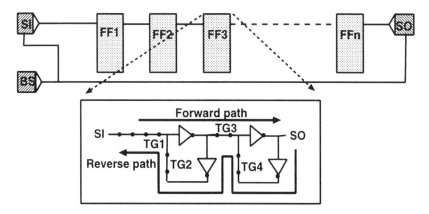

Fig. 4.8: Bidirectional transparent scan to detect open defects.

4.4.2 Bidirectional Transparent Scan

There is little evidence that open defects constitute a significant portion of the total defect population in a CMOS process [26,29]. In two separate studies, Woodhall et al. [29] and Turner et al. [26], found little or no evidence of a purely stuck-open behavior in ICs. Nevertheless, the proposed method will also detect most of the open defects in the scan chain by voltage testing. Open defects in feedback paths of the flip-flop (Fig. 4.3(a), nodes m3 and $\overline{Q}$) will not be detected. If in a given process, the probability of open defects is significantly high, then the transparent scan should be modified to make it bi-directional. In the reverse path shown in Fig. 4.8, open defects in feedback paths will be detected. This will require bi-directional SI and SO ports and a control signal, BS, to switch the direction of the scan path. A logic low on BS selects the forward path and the logic high selects the reverse path. Additional two vectors are needed to test the open defects that are otherwise not detected.

4.4.3 Transparent Scan with Single Clock

The basic idea is to exploit the normal mode non-clock signals (e.g., TC/SE) as clock signals in the test mode to reduce test and implementation complexi-

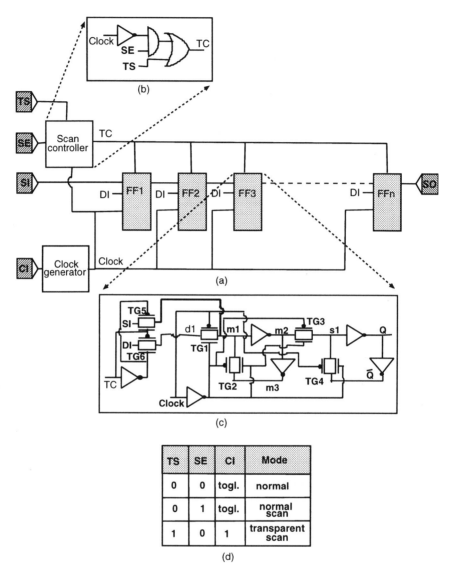

Fig. 4.9: (a) Transparent scan with single clock, (b) scan controller logic diagram, (c) modified scan flip-flop to facilitate single clock operation, and (d) modes of flip-flop operation.

ty. Since these are non-clocked signals in the normal mode, they are not timing critical. Moreover, normal mode clock signal is kept unchanged so that normal mode performance is not affected. It can be argued that this is a better, cost effective approach compared to building additional control for the normal mode clock signal [19,20].

An implementation of the idea is shown in Fig. 4.9. Fig. 4.9(a) illustrates a scan chain, clock generator and an additional block, scan controller, that controls the modes of operation of flip-flops in the scan chain. Fig. 4.9(b) depicts the logic schematic of the scan controller. This implementation requires 3 logic gates. Fig. 4.9(c) illustrates the logic schematic of the scan flip-flip. The flip-flop requires no extra logic gates. Only modification needed in the flip-flop is to reroute the front-end multiplexer outputs. The output of TG5 which normally goes to input of TG1, is now connected to the output of TG2. This change is shown by a bold line in Fig. 4.9(c). The normal mode data path of the scan flip-flop is not changed to keep the normal mode set-up and hold times of the flip-flop unchanged. Furthermore, the clock path of the flip-flop is also not changed. Thus the normal mode operation of the flip-flop also remains unaffected. Essentially, in the scan mode, TG1 is replaced by TG5, which is clocked by TC. Rest of the TGs (TG2-TG4) are clocked by the clock. The signal TC can be controlled asynchronously at logic high as well as logic low. Moreover, it can act as a clock, depending upon the input decoding conditions. These conditions are shown in Fig. 4.9(d). In the transparent scan mode, TC is kept at logic high by virtue of logic high on input TS. This ensures that TG5 is conducting. Clock is held at logic high so that TG2 and TG3 are conducting. Hence, the flip-flop becomes transparent in this mode. Therefore, the complete scan chain can be quickly tested.

Mercer and Agrawal [13] were one of the first to investigated optimization of clock and scan enable signals. Their motivation was to save routing of an extra signal to two-phase clock flip-flops. In a two-phase clock scan flip-flop, two clock signals, and a scan enable signal are typically needed. These control signals, can have $2^3 = 8$ control states. However, not all control states are used in a scan flip-flop, and some control states can be decoded locally. A simple decoding logic can be a part of a flip-flop. As a result, only two control signals are required for each flip-flop. Alternatively, similar to the clock generator modification discussed above, clock generator may be modified to reduce the

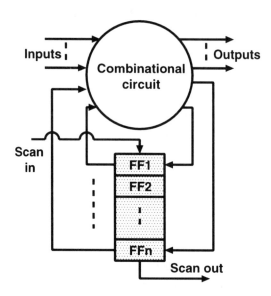

Fig. 4.10: Usage of scan chain to avoid the oscillatory feedback loops.

area required for local decoding of signals. In their scheme, the scan chain transparency (flush test) is not possible.

Subsequently, Bhavsar implemented this technique with very few MOS transistors [2]. He proposed a modification of latches in single latch design style of LSSD such that in scan mode the latches function as dynamic master-slave flip-flops. This technique requires very little area overhead for scan implementation in single latch design style. However, it has some disadvantages. The scan chains are dynamic in nature, therefore, extra care must be taken while propagating data through them. Secondly, the flush test is not possible in this methodology. Finally, in this technique, a static latch has been converted into a dynamic master/slave flip-flop. As a result, the feedback paths are turned off during the shift test. Therefore, testing defects in the feedback paths require an additional hold test.

4.4.4 Why Scan?

Fig. 4.10 illustrates the Huffman model of a digital circuit. In this model, combinational feedback loops are broken by the sequential logic such that oscillatory states do not invalidate the circuit function. When flip-flops are made transparent, combinational feedback will become active. Therefore, the flip-flop transparency may give rise to oscillating feedback loops that will invalidate QCM and voltage based tests. A cost effective solution is to put all flip-flops in a scan chain and then make this scan chain transparent. It may be recalled that the flush test of Bottorff et al. [33] is also applied only in the scan mode. In this configuration, feedback loops are broken by the scan path and therefore the tests are not invalidated. Since a complete scan chain in the test mode (transparent scan) contains only inverters and is effectively controlled/ observed by primary inputs and outputs. It is easily tested by voltage as well as QCM tests compared to testing of individual transparent flip-flop in isolation. Moreover, the test complexity of the transparent scan chain is independent of the number of flip-flops in it.

4.4.5 Implementation Aspects

The chief requirement for the implementation of the transparent scan is an independent control of Clock and $\overline{\text{Clock}}$ signals. Both signals should be available as outputs of the clock generator. The modified clock generator contains a set of extra drivers for $\overline{\text{Clock}}$ line and a 2-input OR-gate in the $\overline{\text{Clock}}$ generation path to provide an additional control over $\overline{\text{Clock}}$. To make the scan chain transparent, Clock and $\overline{\text{Clock}}$, both should be held at logic low. Clock is controlled from the primary input CI while $\overline{\text{Clock}}$ is controlled by another primary input TS. Each individual flip-flop must contain an extra inverter to obtain the complement of $\overline{\text{Clock}}$ signal. A flip-flop implementation is shown in Fig. 4.6(c). Furthermore, Clock and $\overline{\text{Clock}}$ lines must be routed instead of only the Clock line. Changes in flip-flop as well as in the Clock generator result in a marginal increase in the IC area.

The impact of the proposed method on Clock and flip-flop timing is insignificant. The Setup and Hold times of a flip-flop are governed by the master-latch. TG2 now gets clock through $\overline{\text{Clock}}$ which should not affect the timing. Similarly, the propagation delay through the flip-flop should not be changed in any significant way. The clock driver can be carefully designed to minimize any timing impact. On the other hand, the design of the clock driver may be simplified, since each of the two parallel branches drives precisely half of the capacitive load. The global solution becomes even more attractive for flip-flops with two-phase clock. Since, both phases of the clock are routed on separate wires, only additional implementation requirement is the independent control of the clock phases. This could be done in the same manner as explained for single phase clocked ICs.

The single clock implementation of the transparent scan saves the routing of an extra clock signal in the layout. It utilizes a non-clock signal in the test mode. Therefore, the impact of its implementation on timing and area is arguably minimal. However, in this case, bidirectional nature of the scan chain can not be exploited. As mentioned before, since the open defects are significantly less probable, this implementation can be exploited without sacrificing the quality of the tested devices.

The potential of this concept is substantial in terms of scan chain test complexity reduction and test quality improvement. First of all, the sequential test complexity is broken down to a level where the test consists of four or six test vectors depending on whether a unidirectional or a bidirectional transparent scan is implemented. The test complexity of combinational logic is lower and it can still be tested with the combination of transparent scan and normal scan test. The reduction of total chip complexity by transparent scan is a subject for further research. Secondly, this method ensures that the bridging defects in sequential circuits that are otherwise either hard to detect or are not detectable. Thirdly, complex ICs often utilize multiple scan chains because a large number of sequential elements in one scan chain elevates its test complexity. The test complexity of a scan chain is proportional to the number of elements in it. In a relatively long scan chain, it takes much longer to shift data in and out. On the other hand, multiple scan chains may require larger number of dedicated or multiplexed test pins. The test complexity of the scan chain with the proposed method is independent of the number of elements. Therefore, one scan chain can be utilized. On the other hand, one extra input pin will be required for providing control over the clock line. Another input pin will be needed if bi-directional transparent scan mode is desired. Similar to the normal scan inputs, these pins can be multiplexed with the normal input pins. Although the global solution is attractive at chip level, in some situations, a clever utilization of a local solution can also improve the local controllability and observability of the combinational logic.

4.5 Testable Flip-flop Configurations

Although the scan path transparency significantly reduces the voltage test complexity of scan paths and substantially improves the fault coverage of the I_{DDQ} test, its application requires the presence of a scan path in the IC. According to the projections of the semiconductor industry association (SIA), the majority of ICs will continue to be produced without any DfT or scan path [22]. Furthermore, for scan path transparency, an independent control of Clock and $\overline{\text{Clock}}$ signals is required. Therefore, an additional global signal ($\overline{\text{Clock}}$) must be routed to achieve the transparency. Routing of an extra ($\overline{\text{Clock}}$) signal adds additional timing complexity in the design. A proper timing relationship between Clock and $\overline{\text{Clock}}$ must be ensured over the entire IC for a proper operation. Close proximity of high speed Clock and $\overline{\text{Clock}}$ lines for extended length may give rise to cross-talk, noise, etc. Routing of an additional global signal may cost additional area.

Considering these aspects, there is a strong motivation to find alternative design/test techniques for I_{DDQ} detection of such bridging faults in sequential circuits. Therefore, two new flip-flop configurations have been proposed with following characteristics: (i) Bridging faults are I_{DDQ} detectable. The I_{DDQ} detection of a bridging faults is no longer constrained by design level parameters with respect to the fault resistance. Hence, bridging fault detection is solely determined by the I_{DDQ} threshold. (ii) The voltage detection of faults remains unchanged.

We shall demonstrate that owing to the use of transmission gates in CMOS flip-flops, some of the bridging faults are not detected by I_{DDQ} test method. The proposed configurations solve this problem. Design considerations are

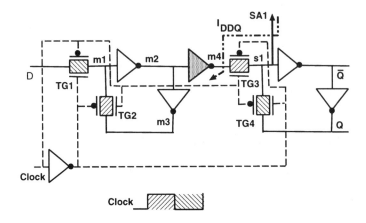

Fig. 4.11: A schematic of first testable flip-flop configuration.

presented in Section 4.5.1 and Section 4.5.2. Section 4.5.3 is devoted to a bridging fault detection strategy. Four I_{DDQ} samples are needed to detect bridging faults. All flip-flops in the DUT can be tested in parallel with the I_{DDQ} test technique. In high performance applications, flip-flop timing performance is a more important issue than its I_{DDQ} testability. Section 4.5.4 addresses the timing characterization of the proposed flip-flop configurations and compares them with the conventional configuration. Simulation as well as analytical results show that proposed flip-flop configurations compare favorably with the conventional flip-flop configuration.

4.5.1 First Flip-flop Configuration

Shorts or bridging faults in the slave latch can be made I_{DDQ} detectable if the master latch is prevented from being over-written due to a fault in the slave latch. Since the master latch is not over-written, the data stored in master sustains a logical conflict with the fault in the slave. For example, a logical conflict is created and sustained between the fault and node s1 while TG3 is conducting (Fig. 4.11). The prevention of master latch over-writing is achieved if TG3 is made unidirectional such that it transfers data only from the master to the slave latch. TG3 is made unidirectional either by putting an additional inverter just before it or by replacing TG3 by a clocked-inverter. A

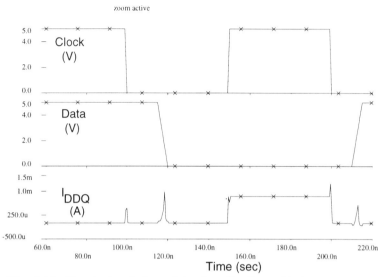

Fig. 4.12: Circuit simulation of the first flip-flop with fault showing the detection.

schematic of the flip-flop configuration is shown in Fig. 4.11. The additional inverter improves fault detection by I_{DDQ} test method and also improves various timing aspects of the flip-flop. Since the master latch is effectively isolated from the slave latch, setup and hold times of the flip-flop are improved. These issues will be examined in subsequent sections in greater detail.

Let us again consider the bridging fault (or SA1 fault) at node s1 (Fig. 4.1) once again. The input data of the flip-flop is kept at logic low (therefore, node m4 is low) while the Clock is also low. Subsequently, the Clock is switched to logic high. The change in the Clock state causes TG3 to start conducting. As a result, a logical conflict is created between node m4 (driven low by the preceding inverter) and node s1 (driven high by the bridging fault). This conflict is sustained so long as the Clock stays in logic high state. The path of the current flow is illustrated in Fig. 4.11. In this case, the n-channel transistor of the additional inverter sustains the conflict with the fault. Fig. 4.12 illustrates the result of the Pstar (an in-house circuit simulator at Philips) simulation on the flip-flop with a low resistive bridging fault. The simulation result shows that while Clock is logic low no I_{DDQ} flows. However, elevated I_{DDQ} flows so long

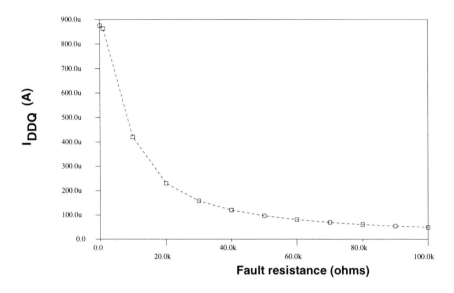

Fig. 4.13: A plot of I_{DDQ} as a function of fault resistance.

as the Clock is logic high (TG3 is conducting) and previously stored data in the master latch is logic low. Approximately, 0.9 mA of quiescent current flows through the fault if the resistance of the fault is small.

A plot of I_{DDQ} as a function of the bridging fault resistance is shown in Fig. 4.13. At the instance when the fault resistance is zero, approximately 0.9 mA current flows. Since the fault resistance is zero, all the voltage is across the n-channel transistor during the conflict. Therefore, the on-resistance, R_{on}, of the conflicting transistor is computed approximately as 5.5k ohms. As the resistance of the fault rises, the I_{DDQ} is largely determined by the combined resistance of the bridging fault and on-resistance of the conflicting transistor. For example, when the fault resistance is 10k ohms, approximately 0.35 mA current should flow. However, larger current flows due to the fact that intermediate voltage levels on nodes m4 and s1 cause additional quiescent current in the inverters of the slave latch. Finally, when R_{def} becomes considerably larger (5 times, or more) than R_{on}, the I_{DDQ} is solely determined by the fault resistance and I_{DDQ} contribution by other inverters becomes negligible. The Pstar simulation result are shown in Fig. 4.13. A monotonic decrease in I_{DDQ} with increasing fault resistance suggests that I_{DDQ} detection of the fault does not

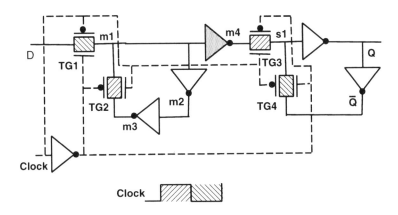

Fig. 4.14: A schematic of second flip-flop configuration.

depend on the circuit level parameters. A similar graph for a conventional flip-flop obtained by Rodriguez-Montanes et al. [16] showed a peak in I_{DDQ} when fault resistance was 1.5 kilo ohms. Effectively, it means that for a conventional flip-flop, I_{DDQ} does not detect the fault for all fault resistance values.

4.5.2 Second Flip-flop Configuration

In the above flip-flop, there are four inverters between data input (D) and data output (Q) instead of just two inverters in the conventional flip-flop. This design of the flip-flop will cause an additional propagation delay that may not be acceptable in some circumstances. The propagation delay (t_{pd}) of a flip-flop is defined as the elapsed time between signals Clock and output Q. It is calculated from the instant when the active edge of the Clock reaches VDD/2 to the instant when Q reaches VDD/2. Data is assumed to be stable and propagated till the input (m4) of the slave latch before the active edge of Clock. At the Clock transition, data starts to propagate through the slave latch. Therefore, the propagation delay of a flip-flop is effectively the delay from node m4 to output Q. Higher propagation delay in the proposed flip-flop is caused by the presence of an extra inverter in the slave data path.

In delay critical situations, the above mentioned flip-flop can be slightly modified to substantially reduce the propagation delay. This flip-flop configuration is shown in Fig. 4.14. In the master latch, two inverters are put in the feedback path instead of one in the previous design. As a result, in the slave latch, the Q output is taken one inverter earlier. Effectively, there are only two inverters between input (D) and output (Q). The extra inverter no longer contributes to the propagation delay of the flip-flop. We shall look into flip-flop timing issues in Section 4.5.4. Since, the master latch remains isolated from the influence of the slave latch, this flip-flop is also I_{DDQ} testable.

4.5.3 Fault Detection Strategy

The detection of bridging faults in a flip-flop can be explained as follows: In CMOS circuits a state is memorized by a latch. Lee [11] described it as a control loop. A control loop is in the floating state if it retains the previous value forced by its driver, otherwise, it is in the forced state (i.e., being driven or written). Due to the presence of floating control loops, a bridging fault may escape detection. A bridging fault in a control loop can only be detected when the loop is in the forced state. Master and slave latches form two individual control loops. The master latch can be put into the forced state while TG1 is conducting. However, in conventional flip-flop configurations, the slave latch can never be put into the forced state. It is rather coached into memorizing the state by the master latch. During the period of coaching (TG3 is conducting) the master latch is in the floating state. Therefore, if the slave latch offers a large inertia (e.g., in the presence of a bridging fault) against the state change, the master latch itself is over-written. In the proposed flip-flop configuration we replace this coaching by explicit forcing when TG3 is conducting. Therefore, inertia can no longer over-write the master latch.

4.5.3.1 I_{DDQ} Testing

Bridging faults in sequential circuits are put into three categories depending upon the number of floating loops that are involved [18]:

1. Bridges influencing nodes not belonging to floating control loops.
2. Bridges with only one node belonging to a floating control loop.
3. Bridges with two nodes belonging to two different floating control loops.

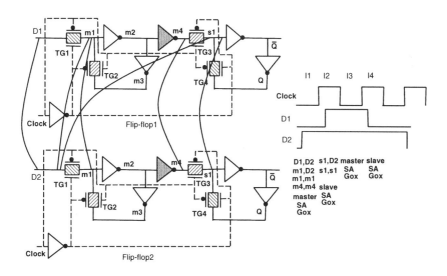

Fig. 4.15: Classes of bridging faults in sequential circuits and their I_{DDQ} detection strategy.

In Fig. 4.15, we outline the fault detection strategy with examples of 6 bridging faults between two proposed flip-flops. These faults are representatives of the above mentioned three categories. For example, the bridging fault (D1, D2) is representative of the category 1. Similarly, bridging faults (m1, D2) and (s1, D2) are representatives of the category 2 and bridging faults (m1, m1), (m4, m4), and (s1, s1) are representatives of the category 3.

Without the loss of generality, let us assume that flip-flop1 is under investigation. Of the six bridging faults, four affect the master latch and the remaining influence the slave latch of the flip-flop. Let us further assume that input (D2) of flip-flop2 is high for the entire duration of the test. A total of four I_{DDQ} samples (two clock cycles) as indicated in Fig. 4.15, are needed to test all bridging faults including the above mentioned faults in flip-flop1. The first I_{DDQ} measurement, I1, tests the bridging faults (D1, D2), (m1, D2), (m1, m1), (m4, m4), and some gate oxide defects and SA faults in the master latch. The second I_{DDQ} measurement, I2, detects bridging faults (s1, m1), (s1, s1), and some gate oxide defects and SA faults in the slave latch. Now, consider node m2 has a bridging fault to VSS or n-channel transistor of the first inverter in master latch has a gate oxide defect to VSS. These defects/faults will be detected by I3 I_{DDQ} measurement when D1 is kept high. Similarly, I4 measurement will detect remaining gate oxide defects and bridging faults in the slave latch that are not detected by I2 measurement.

In the situation where flip-flops form structures like a shift register, data transfer from a flip-flop to a succeeding flip-flop should be made unidirectional so that bridging faults in the master-latch of the succeeding flip-flop are also detected by I_{DDQ} testing. This situation is different from flip-flops in a scan path where in the case of a bridging fault in the master-latch, a logical conflict across the fault can be created in the non-scan (normal) mode. Similarly, in the situations where latches are used without combinational logic in between, the data transfer between them can be made unidirectional to preserve the I_{DDQ} testability of faults.

4.5.3.2 Voltage Testing

The voltage testing of the proposed flip-flop does not require explanation. Two Clock cycles with data low and high are needed to test a flip-flop by voltage measurements. The Boolean operation of the proposed flip-flop remains the same, therefore, the proposed flip-flops can also be tested in two clock cycles. Voltage testing will detect low resistive bridging as well as open faults.

4.5.3.3 Undetected Bridging Faults

In the proposed flip-flop configurations, some bridging faults belonging to category 3 will not be detected by the I_{DDQ} test method. For example, bridging fault (m2, s1) will not be detected. In such cases, at least one latch is in the floating state, and the logical conflict is not created. These faults can be detected by I_{DDQ} testing, if clocks going to both of the affected flip-flops can be independently controlled. Majority of bridging faults in a flip-flop affect nodes within the flip-flop (categories 1 and 2). The probability of occurrence of such faults belonging to category 3 is significantly small. The probability can further be reduced by proper layout and logical considerations.

4.5.4 Timing Characterization

Timing performance of a flip-flop is characterized by data setup time (t_{su}), hold time (t_h) and the propagation delay. The setup time is defined as the time before which the data input should be stable with respect to the edge of the clock. Similarly, the hold time is defined as the time after which the data should be stable with respect to the edge of the clock. The propagation delay was explained earlier. For determination of the setup time, a given flip-flop is initially simulated with relaxed setup time. Subsequently. data is changed successively closer to the active edge of the clock while the output of the flip-flop is kept under observation. At the instance when the output of the flip-flop fails to register the change from input data, the time difference between input data and the clock edge is considered to be its setup time. This time difference is calculated from midpoints (VDD/2) of these signals. The hold time of a given flip-flop is also calculated similarly. Initially, the flip-flop is simulated with relaxed hold time. Subsequently, after the active edge of the clock, data is changed successively closer to the active edge while the output of the flip-flop is observed. At the instance when the output of the flip-flop fails to register the change in input data, the time difference between the clock edge and data edge is considered to be its hold time.

To compare performances of the conventional and proposed flip-flops, we consider setup time, hold time and propagation delays. For this comparative study, it is ensured that all flip-flops have the same transistor dimensions and only an extra inverter of similar dimension is added in the proposed flip-flops. Transistor sizes are selected to be representative of the technology and the design style. For this analysis, we selected a standard 1.0 micron single poly, double metal technology. However, no extra effort is made to particularly optimize flip-flops for power, performance, area, etc. All flip-flop configurations are simulated with the circuit simulator, Pstar. All simulations were carried out at nominal voltage and temperature conditions with typical process param-

Table 4.1: Setup, hold times and the propagation delays of the conventional and proposed flip-flops.

Timing	Flip-flop					
	Conventional		Proposed1		Proposed2	
	Data 1	Data 0	Data 1	Data 0	Data 1	Data 0
Setup time (ns)	0.48	0.53	0.39	0.39	0.36	0.38
Hold time (ns)	-0.30	-0.31	-0.38	-0.39	-0.36	-0.38
Prop.delay (ns)	0.42	0.46	0.78	0.68	0.56	0.59
Setup + Prop.delay	0.90	0.99	1.17	1.07	0.92	0.97

eters (level 9 MOST model). We must note that the selection of absolute transistor parameters or flip-flop optimization for performance was of minor consequences in this study. It is the relative performance of the flip-flop configurations that is evaluated to quantify the impact of proposed flip-flop configurations on the timing performance.

4.5.4.1 Setup and Hold Times

In Table 4.1 a comparison of timing is shown. The setup time of the first flip-flop (proposed 1) is less than the setup time of the conventional flip-flop. For data 1, this flip-flop has 0.39ns as the setup time compared to 0.48ns for the conventional flip-flop. The setup time of the second flip-flop (proposed 2) is fractionally lower than the setup time of the first flip-flop. Similarly, hold times of the proposed flip-flops are considerably lower compared to that of the conventional flip-flop. For example for data 1, the hold time of the first flip-flop is -0.38 ns compared to -0.30 ns of the conventional flip-flop. Conventionally, the hold time is defined as the elapsed time after the clock edge for which

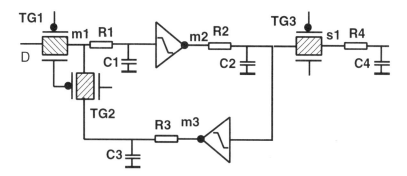

Fig. 4.16: Simplified master latch model of the conventional flip-flop.

data should be stable. However, it is not uncommon to have negative hold time for a flip-flop. The negative hold time signifies that the flip-flop is able to latch-in the data even if data changes before the clock edge. Lower hold time is a desirable feature for a flip-flop.

Smaller setup and hold times of proposed flip-flops can be explained with the help of a simplified, first order model of the master latch (Fig. 4.16) of the conventional flip-flop. This model consists of two static logic gates (representing the non-linearity of the flip-flop) and a linear dynamic network (R1,C1,....R4,C4) describing its dynamic behavior [9]. In this model, we have also included the input resistance and capacitance of the slave latch since it influences the flip-flop setup and hold times. A TG used in the conventional flip-flop offers three times the capacitive loading compared to a simple inverter. Therefore, loading not only D input but also internal nodes [9]. Furthermore, owing to the bidirectional nature of TG3, a part of the slave latch inertia against state change is passed on to the master latch during the 0->1 transition of the clock. The extra inverter in proposed flip-flops (or clocked inverter instead of TG) provides a special buffering and effectively reduces the loading of node m2. As a result, the RC time constant is reduced and the gain-bandwidth product of the master latch is improved for fixed transistor dimensions and technology [9,27]. Therefore, setup and hold times of proposed flip-flops are approximately 25% lower compared to the conventional flip-flop.

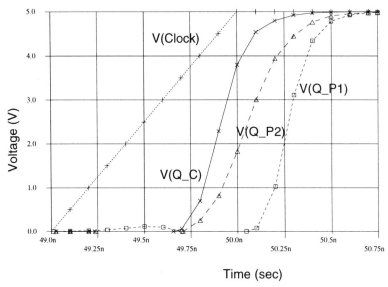

Time (sec)

Fig. 4.17: A plot of propagation delays of the conventional and proposed flip-flops.

4.5.4.2 Propagation Delay

As mentioned earlier, the first of the proposed flip-flops has considerably larger propagation delay (0.78ns for data 1) compared to the propagation delay of the conventional flip-flop (0.42ns) at no load condition. The extra propagation delay largely comes from the extra inverter in the data path. The second flip-flop eliminates the extra inverter delay, however, the propagation delay still remains significantly higher (0.56ns). The extra propagation delay is compensated by lower setup time and the maximum toggle frequency of the second flip-flop is comparable to that of the conventional flip-flop. Fig. 4.17 graphically shows the relative propagation delay of three flip-flops at no load condition. The output of the conventional flip-flop, Q_C, shows the lowest propagation delay. The output of the first flip-flop, Q_P1, shows the highest propagation delay and the output of second flip-flop, Q_P2, shows propagation delay in between the two.

In spite of the same data path (D-->Q), the conventional flip-flop has a lower propagation delay compared to that of the second flip-flop. The difference in

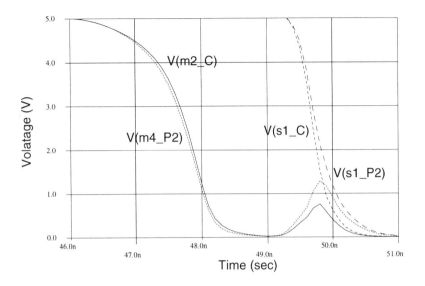

Fig. 4.18: A plot of the internal node voltages of the conventional and
second proposed flip-flops.

their propagation delays is explained with the help of Fig. 4.18. This figure
shows the result of Pstar transient analysis at internal nodes of both flip-flops.
The voltage on nodes m2_C and s1_C for the conventional flip-flop and corresponding nodes (m4_P2 and s1_P2) for the second flip-flop are plotted. The
"C" stands for conventional flip-flop and the "P2" signifies the second proposed flip-flop. As can be seen from the plot nodes m2_C and m4_P2 are fairly close to each other except at the instance of clock transition where both
nodes show visible bounces of different magnitude. However, the node m2_C,
being a part of the regenerative positive feedback loop, manages to settle
down quickly as compared to the node m4_P2. This small time difference becomes larger at nodes s1_C and s1_P2 and eventually causes different propagation delays.

4.5.4.3 Metastable Behavior

The gain-bandwidth product of a flip-flop is a measure of how quickly the re-generative configurations of latches and flip-flops recover from the metastable state. Various parameters (e.g., threshold voltage, transistor aspect ratios, substrate doping [4,9,10,27]) can be optimized to improve the gain-bandwidth product of a flip-flop. As discussed earlier, the gain-bandwidth product of a flip-flop with constrained transistor sizes is improved by reducing the RC time constant affecting the regenerative feedback loop. The parasitic capacitance of the master loop in the proposed configurations is lower for constrained transistor sizes. Therefore, the metastable behavior of the proposed flip-flops is expected to be better than that of the conventional flip-flop. Improvements in flip-flop and latch metastability have been reported with the usage of clocked inverters instead of TGs [9,27], however, due to their larger transistor count and higher propagation delay (reduced maximum toggle rate) such implementations are not popular with logic designers. The proposed flip-flops provide excellent alternatives as high performance, I_{DDQ} testable flip-flops. The RC time constant of master and slave latches can further be reduced by utilizing only n-channel transistor in TGs instead of p and n-channel transistors. The usage of only an n-channel transistor instead of p and n-channel transistors in TGs will cause neither elevated I_{DDQ} nor any degradation of I_{DDQ} fault coverage in the proposed flip-flops.

The improvement in metastability can also be explained from another perspective. The time-window between setup and hold times ($t_{su} + t_h$), can be termed as the metastability window. The behavior of a flip-flop with respect to the change of data in this time-window is not defined. Change of data in this window may lead to flip-flop metastability. Therefore, the width of this window is a figure of merit for robust, timing insensitive flip-flop. As can be interpreted from Table 4.1, both proposed flip-flop configurations have very small metastability window compared to the conventional flip-flop configuration. The second proposed flip-flop configuration offers the smallest metastability window with a width of less than 1/100 of a nanosecond compared to a width of 0.18 ns for the conventional flip-flop configuration.

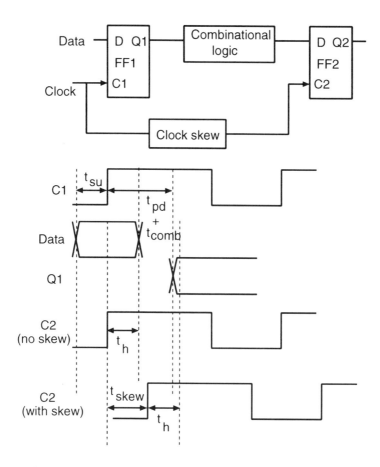

Fig. 4.19: Illustration of the clock-skew robustness of a flip-flop.

4.5.4.4 Other Timing Benefits

The ability of a flip-flop to withstand clock-skew is an important parameter in robust, timing insensitive design. The degree of flip-flop robustness against clock-skew is explained with the help of Fig. 4.19. Consider two similar flip-flops, FF1 and FF2, such that the output of FF1, Q1, through the combinational logic with propagation delay t_{comb}, drives the D input of FF2. For simplicity, let us assume both flip-flops have the same setup and hold times and propagation delays. Let us further assume that clock coming to FF2 has a clock-skew of t_{skew} ns with respect to clock driving FF1. The t_{skew} could be positive as well as negative. In the first instance of C2, there is no clock-skew and the input data for FF2 changes after its hold time. Therefore, correct flip-flop operation is ensured. However, in the second instance of C2, the data for FF2 changes during the hold time and, therefore, a proper operation of FF2 is not ensured. For correct operation of FF2, the maximum possible t_{skew} allowed is given by the following equation:

$$t_{skew} = (t_{pd} + t_{comb}) - t_h \qquad \text{(EQ 4.4)}$$

If t_{comb} is very large, the maximum allowed t_{skew} will also be very large. In other words, the circuit has high clock-skew immunity. However, this is rarely the case. The worst case clock-skew situation occurs when t_{comb} is zero. In such cases Q1 directly feeds to FF2 (e.g., shift register) and EQ. 4.4 reduces to:

$$t_{skew} = t_{pd} - t_h \qquad \text{(EQ 4.5)}$$

High propagation delay and low hold time are desirable flip-flop properties in timing robust design. From data in Table 4.1, it is concluded that the first proposed flip-flop can withstand a clock-skew of 1.17ns compared to 0.73ns of the conventional flip-flop. The second proposed flip-flop can absorb a clock skew of 0.94ns. Therefore, in clock-skew sensitive situations, use of the first proposed flip-flop, and in timing critical paths use of the second proposed flip-flop is desirable. This will ensure timing robustness of the design while maintaining its I_{DDQ} testability. Furthermore, higher the insensitivity of a flip-flop to clock-skew, less stringent are the timing constraints on clock generator and clock routing.

4.6 Conclusion

Flip-flops are indispensable building block in digital ICs. However, most of the present static flip-flop configurations suffer from poor coverage of bridging faults by the I_{DDQ} test technique. Bridging faults are amongst the most prevalent faults in ICs. Considering that a complex digital IC may contain several thousand flip-flops and I_{DDQ} is emerging as the quality improving complement to logic testing, such I_{DDQ} escapes will have a severe quality impact. Researchers have attributed this poor coverage to the architecture of flip-flops where the controllability condition is not sufficient for bridging fault detection with the I_{DDQ} test technique.

In this chapter, several solutions of testing defects in flip-flops and scan chains have been suggested. By virtue of flip-flop and scan path transparency defects can be effectively and efficiently detected [19,20]. Flip-flop or scan path transparency requires an independent control of Clock and $\overline{\text{Clock}}$ signals. We have demonstrated that the test complexity of transparent scan is very small and is independent of the number of flip-flops in the scan chain. A number of options for scan path transparency are illustrated. However, scan path transparency requires extra logic and control pin(s) to transform the existing scan chain in the transparent mode.

In situations where scan path is not present or scan path transparency is not available, we proposed two flip-flop configurations that are I_{DDQ} testable. In these flip-flop configurations, the controllability condition is enough for bridging fault detection with I_{DDQ} test technique. I_{DDQ} testability of flip-flops is achieved by converting the bidirectional data transfer between master and slave latches into a unidirectional transfer such that data is transferred only from the master to the slave latch. An inverter is needed to restrict the bidirectional data flow. Such an approach is superior to scan chain transparency because no control and routing of an extra clock signal is needed. However, a small set of bridging faults is not detected. The likelihood of occurrence of such faults is rare and can be further minimized by layout considerations.

Proposed flip-flop configurations have timing properties that can be exploited for timing robust design. For example, in situations where high clock skew immunity is required, the first flip-flop configuration is preferred because of its high propagation delay and lower hold time. Both configurations have significantly lower setup time that may be advantageous in timing critical paths. The second flip-flop configuration can be used in low propagation delay situations. Furthermore, both proposed flip-flop configurations have better metastable behavior compared to the conventional flip-flop configuration. Finally, the toggle rate of the second proposed flip-flop configuration is comparable to that of the conventional flip-flop. Therefore, the proposed flip-flop can be used for robust, testable circuit design without sacrificing the performance. The proposed flip-flop configurations require an extra inverter to make the data path between the master and the slave latch unidirectional. The increase in the flip-flop area is expected to be between 5-7%.

References

1. K. Baker, A. Bratt, A. Richardson, and A. Welbers, "Development of a CLASS 1 QTAG Monitor," *Proceedings of International Test Conference,* 1994, pp. 213-222.

2. D. Bhavsar, "A New Economical Implementation for Scannable Flip-Flops in MOS," *IEEE Design & Test,* vol. 3, pp. 52-56, June 1986.

3. P.S. Bottorhof, R.E. France, N.H. Garges, E.J. Orosz, "Test Generation for Large Logic Networks," *Proceedings of 14th Design Automation Conference,* 1977, pp. 479-485.

4. H. J. Chao, and C. A. Johnston, "Behavior Analysis of CMOS D Flip-flops," *IEEE Journal of Solid-State Circuits,* vol. 24, no. 5, pp. 1454-1458, October 1989.

5. E.B. Eichelberger and T.W. Williams, "A Logic Design Structure for LSI Testability," *Journal of Design Automation and Fault Tolerant Computing,* vol. 2, no. 2, pp. 165-178, May 1978.

6. F.J. Ferguson and J.P. Shen, "Extraction and Simulation of Realistic CMOS Faults using Inductive Fault Analysis," *Proceedings International Test Conference,* 1988, pp. 475-484.

7. F.J. Ferguson and J.P. Shen, "A CMOS fault extractor for inductive fault analysis," *IEEE Transaction on Computer-Aided Design,* vol. 7, pp. 1182-1194, November 1988.

8. S. Funatsu, N. Wakatsuki and T. Arima, "Test Generation Systems in Japan," *Proceedings of 12th Design Automation Conference*, 1975, pp. 114-122.

9. J. U. Horstmann, H. W. Eichel and R. L. Coates, "Metastability Behavior of CMOS ASIC Flip-Flops in Theory and Test," *IEEE Journal of Solid-State Circuits*, vol. 24, no. 1, pp. 146-157, February 1989.

10. L.S. Kim, and R. W. Dutton, "Metastability of CMOS Latch/Flip-Flop," *IEEE Journal of Solid-State Circuits*, vol. 25, no. 4, pp. 942-951, August 1980.

11. K.J. Lee and M.A. Breuer, "Design and Test Rules for CMOS Circuits to Facilitate I_{DDQ} Testing of Bridging Faults," *IEEE Transactions on Computer-Aided Design*, vol. 11, no.5, pp. 659-669, May 1992.

12. W. Maly, and M. Patyra, "Design of ICs Applying Built-in Current Testing," *Journal of Electronic Testing: Theory and applications*, vol. 3, pp. 397-406, November 1992.

13. M.R. Mercer, and V.D. Agrawal, "A Novel Clocking Technique for VLSI Circuits Testability," *IEEE Journal of Solid State Circuits*, vol. sc-19, no.2, pp. 207-212, April 1984.

14. C. Metra, M. Favalli, P. Olivo, and B. Ricco, "Testing of Resistive Bridging Faults in CMOS Flip-Flop," *Proceedings of European Test Conference*, 1993, pp. 530-531.

15. R. Perry, "I_{DDQ} testing in CMOS digital ASICs," *Journal of Electronic Testing: Theory and applications*, vol. 3, pp. 317-325, November 1992.

16. R. Rodriguez-Montanes, J. Figueras and R. Rubio, "Current vs. Logic Testability of Bridges in Scan Chains," *Proceedings of European Test Conference*, 1993, pp. 392-396.

17. R. Rodriguez-Montanes, E.M.J.G. Bruls and J.Figueras, "Bridging Defects Resistance Measurements in CMOS Process," *Proceeding of International Test Conference*, 1992, pp. 892-899.

18. R. Rodriguez-Montanes and J. Figueras, "Analysis of Bridging Defects in Sequential CMOS Circuits and Their Current Testability," *Proceedings of European Design and Test conference*, 1994. pp. 356-360.

19. M. Sachdev, "Transforming Sequential Logic for Voltage and I_{DDQ} Testing," *Proceedings of European Design and Test Conference*, 1994, pp. 361-365.

20. M. Sachdev, "Testting Defects in Scan Chains", *IEEE Design & Test of Computers*, vol. 12, pp. 45-51, December 1995.

21. M. Sachdev, "I_{DDQ} and Voltage Testable CMOS Flip-flop Configurations," *Proceedings of International Test Conference,* 1995, pp. 534-543.

22. P. Singer, "1995: Looking Down the Road to Quarter-Micron Production," *Semiconductor International*, vol. 18, no. 1, pp. 46-52, January 1995.

23. J.M. Soden and C.F. Hawkins, "Test Considerations for Gate Oxide Shorts in CMOS ICs," *IEEE Design & Test of Computers*, vol. 3, pp. 56-64, August 1986.

24. J.M. Soden, C.F. Hawkins, R.K. Gulati and W. Mao, "I_{DDQ} Testing: A Review," *Journal of Electronic Testing: Theory and applications*, vol. 3, pp. 291-303, November 1992.

25. T.M. Storey and W. Maly, "CMOS bridging faults detection," *Proceedings of International Test Conference*, 1990, pp. 842-851.

26. M.E. Turner, et al., "Testing CMOS VLSI: Tools, Concepts, and Experimental Results", *Proceedings of International Test Conference,* 1985, pp. 322- 328.

27. H.J.M. Veendrick, "The Bahavior of Flip-Flops Used as Synchronizers and Prediction of Their Failure Rate," *IEEE Journal of Solid-State Circuits*, vol. 15, no. 2, pp. 169-176, April 1980.

28. T.W. Williams and K.P. Parker, "Design for Testability--A Survey," *Proceedings of the IEEE*, vol. 71, no. 1, pp. 98-113, January 1983.

29. B.W. Woodhall, B.D. Newman, and A.G. Sammuli, "Empirical Results of Undetected Stuck-open Failures," *Proceedings of International Test Conference*, 1987, pp. 166-170.

CHAPTER 5

Defect Oriented RAM Testing and Current Testable RAMs

RAMs are integral building blocks of modern ICs and systems. As far as the testing is concerned, RAMs suffer from quantitative issues of digital testing along with the qualitative issues of analog testing. This chapter reviews the state of the art of defect oriented testing of RAMs and proposes a RAM test methodology using I_{DDQ} and voltage based march tests. Bridging defects in a RAM matrix, including the gate oxide defects, are detected by four I_{DDQ} measurements. The I_{DDQ} test is then supplemented with voltage based march test to detect the defects (opens and data retention) not detectable by the I_{DDQ} technique. The combined test methodology reduces the algorithmic test complexity substantially.

5.1 Introduction

Semiconductor RAMs probably represent the biggest product segment of semiconductor industry. The intense R&D directed toward RAMs has resulted in several orders of increase in the RAM chip capacity in the last two decades. RAMs have played a significant role in the electronic revolution that pervades our lives. The performance of modern computers, communication networks and systems heavily depends on the ability to store and retrieve massive amounts of data quickly and inexpensively. Furthermore, RAMs have found their way into such diverse applications like aerospace, automobiles, banking and consumer electronics.

The ever increasing demand for higher density RAMs is matched by test quality and reliability expectations. On one hand, development of high density RAMs puts a severe strain on testing. On the other hand, system reliability and economics have forced a merger of RAMs with CPU or DSP cores on the same substrate. This merger has resulted in a dramatic change especially in the case of DRAMs that now must be fabricated using a process developed for standard logic. This leads to new challenges in the design and testing of embedded DRAMs.

Embedded DRAMs are special in many ways. Not only they are almost analog devices, operating in a hostile digital environment but they are also harder to test owing to system-limited controllability and observability. In addition, they must be designed with the layout density reaching the limits for the available technology. Embedded DRAM test problems are compounded by the fact that often embedded memories are tested through scan chains and the test sequence, when expanded into the individual vectors, can easily exceed the tester's maximum pattern depth [5]. High packing density, standard manufacturing process implementation and dynamic nature of operation make embedded DRAMs susceptible to *catastrophic* as well as *non-catastrophic* defects. *Non-catastrophic* or *soft defects*, as they are popularly known, are too numerous to be ignored [3,52]. Manufacturing of single chip systems at present and in the future will be strongly related to the design of embedded memories. However, reliability of such systems will depend on our ability to test them with sufficient defect coverage.

In this chapter, we discuss the impact of defects on RAMs and their test strategies. This chapter is divided into two segments. The first segment begins with a brief overview of the conventional memory fault models and algorithms. Subsequently, defect-oriented fault models for SRAMs and DRAMs are evolved. The DRAM fault model is evolved considering the catastrophic defects as well as abstract coupling faults. Algorithms are developed to cover these fault models. Finally, the fault models are validated with the manufacturing test results and conclusions are drawn.

Owing to the increasing test costs, parallel testing of RAM bits is gaining wide acceptance. The second segment of the chapter begins with an overview of RAM parallel testing. Subsequently, an I_{DDQ} based parallel RAM test methodology is outlined. The I_{DDQ} test mode is created to reduce the RAM test costs as well as to significantly increase the effectiveness of the I_{DDQ} testing. A test procedure is developed for a given SRAM fault model. In this procedure first I_{DDQ} measurements are carried out to detect bulk of the faults. Subsequently, conventional march tests are applied for faults not detectable by the I_{DDQ} test. Finally, design aspects of I_{DDQ} testable RAMs are considered and conclusions are drawn.

5.2 Traditional RAM Fault Models

RAMs require a special treatment as far as their testing is concerned. Test techniques of the digital domain are not sufficient to cover many defect/fault mechanisms that are likely to occur in RAMs. Their special test requirements have been recognized and addressed by several researchers [2,4,8-10,13,14,18,29,32,33-35,42,50,51,54,57]. A variety of test algorithms, ranging from complexity $O(n)$ to $O(n^2)$ have been evolved. On one hand, these algorithms include simple algorithms, like MSCAN, ATS, that cover only stuck-at faults [2,18] and on the other, algorithms covering complex pattern-sensitive faults have been proposed [8-10,13,50]. Van de Goor [11] and Mazumder and Chakraborty [27] give excellent overviews covering theory and practical aspects of semiconductor memory test.

The evolution of RAM test algorithms is closely related to their fault model development. Abadir et al. [1] segregated RAM test algorithms according to their respective capabilities to detect various failures possible in RAMs. They segregated memory fault models (and hence the test algorithms) into three broad categories, listed below according to their order of complexity:

1. Stuck-at fault model

2. Coupling fault model

3. Pattern-sensitivity fault model

Stuck-at Fault Model: Stuck-at faults are often caused by shorts and opens in RAM cells and address decoders. As a result, one or more cells have a fixed logic state that cannot be overwritten by the write operation. In this fault model, it is assumed that SAFs adequately represent the faulty behavior of the given RAM. Simple test procedures like MSCAN [2], ATS [18], MATS [32] were developed to cover stuck-at faults in memories. Nair proved that MATS covers all stuck-at faults in RAMs independently of the address decoder design [32]. The complexity of these algorithms is linear with respect to the number of memory addresses.

Coupling Fault Model: An important type of fault that can cause a RAM to function incorrectly is the cell coupling fault [1]. Coupling faults occur because of the mutual capacitance between cells or the current leakage from one cell to another [35,51]. Savir et al. [42] defined coupling between a pair of cells such that a $0 \rightarrow 1$ transition in one cell causes a $0 \rightarrow 1$ transition in another cell only for some fixed values of other cells in the neighborhood. Let G denote some pattern in other cells of the memory. For example, let g_1 and g_2 be two cells. A pattern such that $g_1 = 1$ and $g_2 = 0$ is denoted by $G = g_1\bar{g_2}$. When G is void, it is called a general 2-coupling fault between cell i and cell j, on which other n-2 cells in the memory have no influence [42]. When the content of G is limited to a single bit it is called 3-coupling faults. Because of the enormous complexity of a fault model for more than 1 bit in G, only 2-coupling and restricted 3-coupling faults are investigated [35].

Galloping 1's and 0's or GALPAT was proposed by Breuer and Friedman [2] to cover coupling faults. The major disadvantage of this algorithm is its length, which is $O(n^2)$. This makes it impractical for large memories [1]. Nair et al. [32] proposed algorithms of complexity 30n and $n + 32n\log_2 n$ to cover coupling faults. Furthermore, Suk and Reddy proposed two algorithms of complexity 14n and 16n to cover all 2-coupling faults with some restrictions.

A detailed account of these algorithms is given in [51]. Papachristou and Sahgal developed two test procedures of complexity 36n and $24n\log_2 n$ [35]. The two procedures put together have a similar capability as that of Nair et al. and GALPAT, but they require shorter test application time.

Pattern Sensitivity Fault Model: A RAM cell is said to suffer from a pattern sensitive fault (PSF) if its content is influenced by a particular pattern in the array. Hayes demonstrated that testing for unrestricted pattern sensitive faults in large RAMs is impractical. He introduced the concept of neighborhood. Rather than considering each write or read operation on a Cell C_i to be capable of affecting or being affected by the state of every cell in M_r, he assumed that these operations can only involve a certain set of cells N_i called neighborhood of C_i [13]. Hayes [14] and Suk and Reddy [50] tackled the problem of single PSFs in two dimensional memories by using a special type of neighborhood called the *tiling neighborhood*. In this scheme, copies of the tile cover the whole memory in such a way that no copies overlap and no part of the memory is left uncovered except at the boundaries of the memory. You and Hayes suggested that due to some physical defects the contents of a cell may become sensitive not only to the contents of cells in its neighborhood but also to those in the same column [57]. To tackle the complexity of pattern-sensitive faults, Franklin et al. [8,9,10] proposed a new fault concept called Row/Column pattern-sensitivity faults. Instead of a cell being sensitive to the contents of the neighborhood, as suggested by Hayes [13,14], and Suk and Reddy [50], they proposed that the content of a particular cell is sensitive to the contents of the cells in its row and column. They defined a model to encompass such faults and demonstrated that tests required to detect such faults must be of the order $N^{3/2}$ [8].

5.3 A Defect based SRAM Fault Model

The fault models (and hence the test algorithms) proposed in the previous section were largely based upon mathematical models of faulty RAM behavior and not on the actual manufacturing defects. As a result certain algorithms are inadequate in representing the actual failure mechanisms in RAMs (like MATS, MSCAN) and others are probably an overkill in complexity (like GALPAT). Therefore, the need for efficient and accurate test algorithms was increasingly felt.

Dekker et al. [4] applied defect oriented analysis, IFA, to evolve a SRAM fault model and test algorithms. These procedures were reasonably short, yet powerful enough to catch defects. The main objectives of their work was to show the feasibility of a fault model and test algorithm development based on actual device defects. The defects are modeled as local disturbances in the layout of an SRAM array and translated into defects in the corresponding schematic. The electrical behavior of each of these defects is analyzed and classified, resulting in a fault model at the SRAM cell level. Efficient test algorithms were developed using the fault model. The fault model as well as the test algorithms were validated by testing $8k \times 8$ SRAM devices and by performing failure analysis.

For the development of a fault model, an SRAM is divided into three blocks:

- Memory array
- Address decoder
- R/W logic

These building blocks are analyzed separately. However, following a hypothesis of Nair et al. [33] that states that defects in the address decoder and R/W logic can be mapped onto equivalent faults in the memory array. Dekker et al. did not perform explicit IFA on these blocks. The layout of an $8k \times 8$ double poly CMOS SRAM was used as a vehicle to perform the fault model study. The schematic of the SRAM cell is shown in Fig. 5.1. This cell contains four transistors and two pull-up resistors. Each resistor are of 100 G ohms and is composed of high resistive polysilicon.

In their study spot defects resulted in following defects in the layout:

- Broken wires

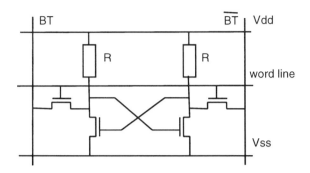

Fig. 5.1 Schematic of the SRAM cell.

- Shorts between wires
- Missing contacts
- Extra contacts
- Newly created transistors

A detailed two-step analysis was carried out for approximately 60 defects. In the first step, defects were placed onto the schematic (Fig. 5.2). In the second step, the defects were classified according to faulty behaviors. Following this analysis, the SRAM fault model had six fault classes:

1. An SRAM cell is stuck-at-0 or stuck-at-1
2. An SRAM cell is stuck-open
3. An SRAM cell has a transition fault
4. An SRAM cell is state coupled to another cell
5. There is a multiple access fault from one SRAM cell to another SRAM cell at another address
6. An SRAM cell suffers from a data retention fault in one of its states. The retention time depends on the leak current to the substrate and the capacitance of the floating node.

Fig. 5.2 illustrates some of these classes of defects in the schematic. For address decoder faults a general fault model was proposed by Nair et al. [33].

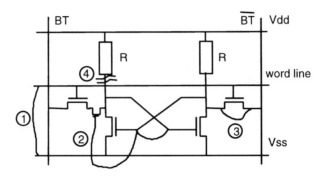

Fig. 5.2: Example of circuit defects for several fault classes.

Under the condition that the faulty decoder stays combinational, the decoder behaves in one of the following manners:

1. A decoder selects more than one addresses
2. A decoder selects no cell

The first situation is equivalent to a multiple access fault and the second situation is equivalent to stuck-open fault in the memory array. Similarly, R/W logic communicates data from I/O ports to the memory array and vice-versa. Faults in busses, sense amplifiers and write buffers result in following fault classes:

1. Data bit(s) having stuck-at fault(s)
2. Data bit(s) having stuck-open fault(s)
3. A pair of bits is state coupled

All these faults in the R/W logic can be mapped as faults in the memory array as well. These three categories of faults are equivalent to a cell stuck-at, a cell stuck-open, and a state coupling faults between two cells at the same address, respectively. Therefore, R/W logic faults are not explicitly considered.

Dekker et al. proposed two SRAM test algorithms of 9N and 13N complexities. The 9N algorithm was developed considering that there is no data latch in

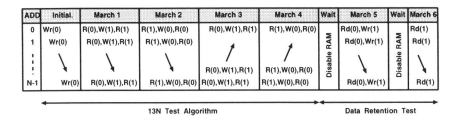

ADD	Initial.	March 1	March 2	March 3	March 4	Wait	March 5	Wait	March 6
0	Wr(0)	R(0),W(1),R(1)	R(1),W(0),R(0)	R(0),W(1),R(1)	R(1),W(0),R(0)		Rd(0),Wr(1)		Rd(1)
1	Wr(0)	R(0),W(1),R(1)	R(1),W(0),R(0)			Disable RAM	Rd(0),Wr(1)	Disable RAM	Rd(1)
⋮				R(0),W(1),R(1)	R(1),W(0),R(0)				
N-1	Wr(0)	R(0),W(1),R(1)	R(1),W(0),R(0)	R(0),W(1),R(1)	R(1),W(0),R(0)		Rd(0),Wr(1)		Rd(1)

13N Test Algorithm   Data Retention Test

Fig. 5.3: 13N SRAM test algorithm proposed by Dekker et al. [19].

the read path. For SRAMs with output data latch the 13N algorithm was developed. The 13N algorithm with the data retention test is illustrated in Fig.
5.3. The effectiveness of these algorithms was validated with $8k \times 8$ SRAMs.
Defective memories were analyzed using microscopes and scanning electron
microscope (SEM) techniques. Validation exercise had twin objectives. The
first objective was to validate the fault model: Do the defects occur in real life
and behave as described in the fault model? The second objective was to validate the effectiveness of the test algorithms compared to test algorithms proposed in the literature with respect to catching realistic defects.

A set of 1192 failed devices was selected from 9 wafers belonging to 3 different batches for test data analysis. The analysis resulted in 15 clusters. The
prominent ones are illustrated in Fig. 5.4. Most of the devices (714) appear to
have SA and total chip failures and 14% of failures were stuck-open. Other
distinguishable failures were data retention faults, multiple access faults, and
state coupling faults. Approximately 10% of analyzed faulty behaviors could
not be explained with IFA based defects and remained unexplained.

Subsequently, the effectiveness of the algorithms was compared with many
other algorithms using 480 devices from the total of 1192 failed devices. For
the devices that suffered from total device failures, SA faults were not considered since such failures can be detected by any algorithm. IFA based 9N and
13N algorithms were found to be better than most of the other test algorithms.

Cluster	#devices	Fault class
0	714	SA & total failure
1	169	stuck-open faults
2	18	multiple access faults
3	9	state coupling faults
4	8	?
5	5	?
6	26	data retention faults
-	-	?
14	2	?

Fig. 5.4: Fault clusters of observed fault classes.

5.4 A Defect based DRAM Fault Model

Dekker et. al demonstrated the effectiveness of IFA with device production results. Promising industrial results on the defect oriented testing stimulated wide ranging interest in this method from academia as well as from the industry. Subsequently, Oberle and Muhmenthaler [34] also used defects as a basis for a DRAM test pattern fault simulator. Unlike Dekker, they derived realistic fault information from DRAM failure analysis. They developed test algorithms that covered those failures and the effectiveness of their tests was verified by the fault simulator.

Dekker et al. [4] and Oberle and Muhmenthaler [34] in their respective approaches used only *hard defects* for fault model development. However, DRAMs are also very sensitive to subtle process variations or soft defects. Sachdev and Verstraelen [39] included soft defects in their IFA of embedded DRAMs. The manufacturing process related defects were divided into hard and soft defect categories. In the analysis, both were separately analyzed and their respective impacts were mapped onto the circuit schematic. In this manner a better and more realistic embedded DRAM fault model was developed. For the *hard defects*, VLASIC [55], a catastrophic defect simulator, was utilized. Defects were sprinkled onto the layout and their impact is mapped onto

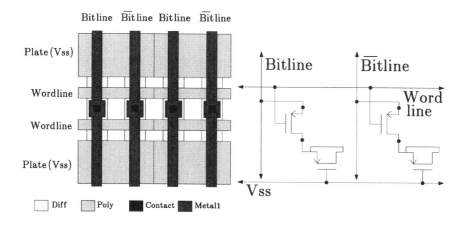

Fig. 5.5: Layout and schematic of a DRAM cell.

the schematic (Fig. 5.5 and Fig. 5.6). Since VLASIC does not handle *soft defects*, a different analytical approach was utilized for such defects. It was assumed that soft defects will cause various 2-coupling faults. The basis of this assumption is that soft spot defects have a local influence that is likely to cause 2-coupling faults. In addition, higher order coupling faults are less likely to occur due to soft defects. Therefore, an exhaustive 2-coupling fault model is developed. Both approaches are explained in the following subsections and a fault model based upon them is developed.

5.4.1 DRAM Cell Architecture

The considered embedded DRAM was realized in a typical CMOS single-poly, double-metal process. The schematic and layout of a memory cell is shown in Fig. 5.5. The core cell consists of two pass transistors and two capacitors to store data and its complement. The two-cell/bit approach was selected for embedded DRAM application on a standard VLSI process because it required less storage capacitance for a given sensitivity of the sense amplifier. The two-

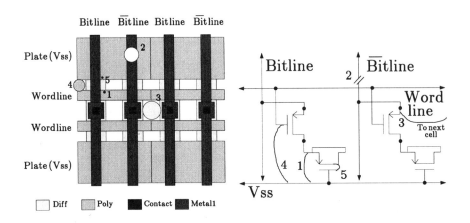

Fig. 5.6: Translation of layout defects onto schematic.

cell/bit approach shows excellent bitline noise rejection in the cell matrix. Furthermore, it resulted in a robust design [48].

5.4.2 Catastrophic Defects

Integrated circuit failures can be attributed to several causes. These causes are broadly divided into global and local disturbances. Global disturbances are primarily caused by defects generated during the manufacturing process. The impact of these global (or manufacturing process related) defects covers a wider area. Hence, they can be detected before functional testing by using simple test structure measurements or supply current tests. A vast majority of faults that remains to be detected during functional testing is caused by local defects, popularly known as *spot defects* [23]. Sachdev and Verstraelen used only spot defects for fault modeling purposes [39]. Spot defects were modeled in the following manner:

For *hard defects*, their analysis was similar to that proposed by Dekker et al. [4]. Defect analysis is performed in two steps. In the first step, defects in the layout are mapped onto the schematic as shown in Fig. 5.6. In the second step, mapped defects are classified in various fault categories. In this manner, contribution of *hard defects* to the fault model is determined. Hard defects resulted in following fault categories:

1. A memory cell is stuck-at-0 or stuck-at-1

2. A memory cell is stuck-open

3. A memory cell is coupled to another cell

4. A memory cell has multiple access faults

5. A memory cell suffers from a data retention fault in one (or both) of its states

Fig. 5.6 shows the translation of physical defects into a circuit schematic. For example, defect 1 is caused by a gate oxide pin hole. Under the influence of this defect the storage capacitor is shorted to the wordline (poly). When the particular cell is accessed, the voltage on the corresponding wordline is always low, making it effectively a stuck-at-0 fault. Defect 2 is an absence of metal1 and causes an open in the bitline. Due to this defect the cell behaves as stuck-open. Extra diffusion, defect 3, causes coupling between two adjacent cells and hence results in a coupling fault. Extra poly, defect 4, causes the poly wordline to be shorted to plate poly (VSS). With this defect, the wordline is always activated and results in a multiple access fault. A high resistance gate oxide pin hole, defect 5, can cause data on the storage capacitor to leak at a rate faster than stipulated and hence results in a data retention fault.

5.4.3 Non-Catastrophic Defects

In the foregoing subsection, the impact of *hard defects* was demonstrated on the circuit and resultant faulty behaviors are explained. In this subsection, the influence of *soft defects* on the development of a fault model is investigated.

Most defects are too small to change the connectivity or logic function of a circuit [52]. On the one hand, such defects degrade the circuit performance and, on the other hand, they can increase the mutual capacitance between adjacent cells or cause current to flow from one cell to another. In other words, they may cause potential coupling faults [35,51]. Furthermore, Bruls [3] high-lighted the potential reliability problems resulting from such defects. As mentioned earlier, owing to their nature of operation, DRAMs, are much more susceptible to such defects than SRAMs or logic circuits. Therefore, coupling faults should be carefully investigated for DRAMs. Abadir and Reghbati [1] defined coupling between two cells as:

A pair of memory cells, i and j, are said to be **coupled** *if a transition from x to y in one cell of the pair, say cell i, changes the state of the other cell j, from 0 to 1 or from 1 to 0.*

Other investigators, e.g., Nair et al. [33] and Suk and Reddy [51], defined coupling faults in RAMs in a similar fashion. As explained in previous section, Savir et al. [42] also gave a comprehensive definition of coupling faults based upon transitions in the coupling cell. According to these definitions, it is the transition in the coupling cell, say i, which initiates the coupling fault. Dekker et. al. [4] defined the concept of *state coupling* for SRAMs, signifying the importance of the state of the coupling cell rather than its transition. He defined state coupling as:

A memory cell, say cell i, is said to be **state coupled** *to another memory cell, say j, if cell i is fixed at a certain value* $x(x\varepsilon\{0,1\})$ *only if cell j is in one defined state* $y(y\varepsilon\{0,1\})$. *State coupling is a non symmetrical relation.*

Clearly, in their definition it is the state of the coupling cell that introduces and maintains the coupling fault in the coupled cell. The important difference between the two definitions is explained as follows: According to the former definition, the coupling is introduced into the coupled cell (j) at the time of transition in the coupling cell (i). Thus, any subsequent *write* operation on cell j will overwrite the coupling fault and it will take another similar transition in cell i to introduce the coupling fault into cell j again. However, according to the latter definition, as long as the coupling cell, say i, is in a particular state, the coupled cell, say j, is also in a particular state and a *write* operation on cell

j should not be able to modify its content. In other words, a particular state of
the coupling cell, *i*, causes a SAF in the coupled cell, *j*.

Before attempting to map either of these definitions onto DRAMs, an important difference between DRAMs and SRAMs should be brought out. In a
DRAM, a cell is driven only when it is accessed and it remains undriven when
not accessed. While performing a *write* or a *refresh*, a cell is driven by the bitline driver. A *read* operation, destroys the cell content, hence, the value is
written back into the cell. Thus, a *read, write* or *refresh* operation on a cell
causes it to be driven. Therefore, at the time of access, the coupling cell, say *i*,
is driven and the coupled cell, say *j*, is not driven. This situation is different
from that of a SRAM. In a SRAM at the time of the coupling fault both the
coupling cell as well as the coupled cell are driven. This special nature of
DRAM operation has a twofold impact on nature of coupling faults:

- The coupling cell *i* has the stronger capability of introducing the coupling
  in cell *j* only when it is in driven state. In the cases where cell *i* is not driv-
  en, it will have a marginal ability of causing the coupling fault in cell *j*.
- The coupled cell *j* is very vulnerable to coupling when it is not driven and
  coupling cell *i* is being driven (accessed).

From this analysis it appears that the definition of coupling based on transitions is enough for DRAMs. But a closer inspection reveals that this is not the
case. In DRAMs, a refresh on cell *i* would re-initiate the coupling in cell *j*, but
it would not change the contents of cell *i*. Consequently, a transition based
definition of coupling does not represent actual coupling in DRAMs. Looking
at Dekker's definition of the state coupling, let us assume that there exists a
coupling between cells *i* and *j* such that logic 1 in *i* forces logic 0 in *j*. A
write(1) on *i* will initiate logic 0 in *j*. Now, *write*(1) is also performed on *j*.
Thus, the coupling is lost. It is assumed that cell *j* will maintain logic 1 despite
the fact that cell *i* is still logic 1. It is based upon the fact that at that moment
cell *i* is not driven and hence is not capable of introducing the coupling into
cell *j* again. However, a refresh on cell *i* can now reintroduce the coupling into
cell *j*. Moreover, as explained above, a *read* operation on a cell in DRAMs is

Table 5.1: Possible 2-coupling faults between cell i and cell j.

Coupling fault	Nature of Coupling
1	(i=0) → (j=0)
2	(i=0) → (j=1)
3	(i=1) → (j=0)
4	(i=1) → (j=1)
5	(j=0) → (i=0)
6	(j=0) → (i=1)
7	(j=1) → (i=0)
8	(j=1) → (i=1)
9	(i=x) → (j=x)
10	(i=x) → (j=$\bar{x}$)
11	(j=x) → (i=x)
12	(j=x) → (i=$\bar{x}$)
13	(i=0) ≡ (j=0)
14	(i=0) ≡ (j=1)
15	(i=1) ≡ (j=0)
16	(i=1) ≡ (j=1)
17	i ≡ j
18	i ≡ $\bar{j}$

destructive by nature. This means that the read value is immediately restored (or written) in the respective cell. Consequently, a *read* operation is also capable of introducing the coupling. Thus, neither definition is suitable for DRAMs. The coupling in DRAMs is dynamic because it is initiated only when the coupling cell, say i, is in the driven state and coupled cell, say j, is not in the driven state. Furthermore, when cell i is not in the driven state it is not capable of initiating the coupling and at best it can *charge-share* with cell j. In this mechanism the contents of both cells are modified and the fault should be detected. Thus, for DRAMs Sachdev and Verstraelen [39] define *dynamic coupling* as follows:

*Two DRAM cells, say i and j, are assumed to be **dynamically coupled**, if the **driven state** of cell i with value $x(x\varepsilon\{0,1\})$ causes cell j to be in state $y(y\varepsilon\{0,1\})$.*

Here no assumption is made about the symmetrical or asymmetrical nature of the coupling. For two arbitrary cells i and j, all possible couplings are shown in Table 5.1. It is assumed that $i_{address} < j_{address}$ while modeling these coupling faults. The other condition, $i_{address} > j_{address}$, has been taken into account by $j \rightarrow i$ coupling faults. For example, coupling fault 1 occurs when logic 0 in cell i forces a logic 0 in cell j. The coupling fault 9 illustrates that logic value x (0 or 1) in cell i forces the same value in cell j. Thus, it appears that coupling faults 1 or 4 are a subset of coupling fault 9. We shall later see that not to be the case. The detection of coupling fault 9 does not guarantee detection of coupling faults 1 and 4. Therefore, these faults should be separately considered. Coupling fault 13 involves a bidirectional coupling. However, it is different from a bridging fault because it is defined only for logic 0. Coupling fault 17 is a bridging fault and coupling fault 18 is a bidirectional inverting bridge. It is important to note that the definition of coupling is extended to include symmetrical behavior as well.

5.4.4 DRAM Test Algorithms

In this section, we first propose a test algorithm for bit oriented DRAMs with combinational R/W logic. We, then extend the algorithm to cover sequential R/W logic and word-oriented DRAMs. The complexity of the basic algorithm is 8N, where N is the number of memory addresses. A data retention test is added to cover data retention faults. Fig. 5.7 shows the flow of the algorithm. It has an initialization step and a set of four marches. In the data retention test, the DRAM is disabled for data retention time and then accessed. The coverage of the fault model by 8N algorithm is explained below.

Stuck-at faults: It is easy to demonstrate the 100% stuck-at fault coverage of the algorithm. The stuck-at-0 fault in any arbitrary cell i will be detected in march 2. Similarly, stuck-at-1 in cell i will be detected in march 1.

Addr.	Initialization	March1	March2	March3	March4	Wait	March5	Wait	March6
0	Wr(0)	Rd(0)Wr(1)	Rd(1)Wr(0)	Rd(0)Wr(1)	Rd(1)		Rd(1)Wr(0)		Rd(0)
1	Wr(0)	Rd(0)Wr(1)	Rd(1)Wr(0)		Rd(1)		Rd(1)Wr(0)		Rd(0)
2	Wr(0)	Rd(0)Wr(1)	Rd(1)Wr(0)	↗	Rd(1)	Disable DRAM	Rd(1)Wr(0)	Disable DRAM	Rd(0)
⋮				Rd(0)Wr(1) Rd(0)Wr(1)					
N−1	Wr(0)	Rd(0)Wr(1)	Rd(1)Wr(0)	Rd(0)Wr(1)	Rd(1)		Rd(1)Wr(0)		Rd(0)

◄─────────────── 8N Test Algorithm ───────────────►◄─── Data retention Test ───►

Fig. 5.7: The 8N DRAM test algorithm with data retention test sequence.

Stuck-open faults: The detection of stuck-open faults is explained with the help of Fig. 5.8, showing a simplified diagram of the data path from storage cells to the local sense amplifier. C_1 ... C_n are the storage cell capacitances. C_{bl} is the bitline (or $\overline{bit}$ line) capacitance and C_t is the truncated bitline capacitance owing to the stuck-open fault. The bit and $\overline{bit}$ lines are terminated on the local sense amplifier. The bitline equalizer circuit is connected to the other ends of the bit and $\overline{bit}$ lines.

The complete read operation is divided into individual steps. In the first step, the bit and the $\overline{bit}$ line voltages are equalized to an approximate voltage of VDD/2. Subsequently, the selected wordline goes low enabling the storage capacitor and line capacitance to charge-share. The bitline capacitance is approximately 10 times that of cell storage capacitance. Thus, after the charge-sharing, the voltage swing on bitline (or $\overline{bit}$ line) is of the order of 400-500 mV (for a DRAM operating with 5V process). Bit and $\overline{bit}$ lines swing in opposite directions because they charge-share with complementary data. The sense amplifier raises the differential voltage to VDD (or VSS) level.

In the first loop the memory is initialized with logic 0. All stuck-open bits on the faulty bitline fail to initialize. However, truncated bitline capacitance, C_t, gets and maintains a logic 0. Subsequently, in the second loop *read*(0) and *write*(1) operations are carried out. Depending upon the location of the open defect, there are several possibilities. Three typical cases are discussed below:

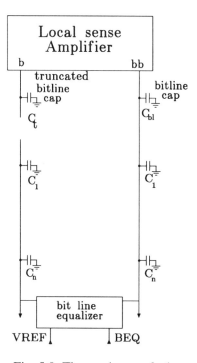

Fig. 5.8: The stuck open fault.

(a) *The stuck-open disconnects all bits from the sense amplifier:* C_t at the beginning of the second loop has logic 0. In the beginning of the first march, *read*(0) is performed. The $\overline{bit}$ line after equalization and charge-sharing swings toward logic 1. However, C_t does not get equalized and does not charge-share with the storage capacitor and hence maintains logic 0. The sense amplifier raises voltage levels on bit and $\overline{bit}$ lines and outputs a logic 0, so the fault is *not detected*. After the *read,* a *write*(1) is carried out, and so C_t obtains logic 1. Moving forward in the same loop, after some time again the same bit and $\overline{bit}$ lines are accessed (let us say location C_2). Now C_t has logic 1. At this moment the truncated bitline has a voltage level of VDD and $\overline{bit}$ has the voltage VDD/2 + 0.4V. This causes the sense amplifier to converge on logic 1 instead of logic 0 and hence the fault will be detected.

(b) *The stuck-open disconnects only the bitline equalizer circuit from the sense amplifier:* In this case, the bitline does not become equalized. However, it charge-shares with the respective cell storage capacitances. It can be shown in the same manner as above that such a defect will be detected.

(c) *The stuck-open disconnects a particular cell from the bitline:* Let us assume that the metal1 to diffusion contact between the bitline and a particular cell, say i, is missing. This defect would cause cell i to be stuck open and hence the cell will not charge-share with the bitline. Thus, at the time of read on this cell, the corresponding bitline has only the equalized voltage. The sense amplifier performance will deteriorate and it will not be able to converge to either logic 1 or 0 in a given time. It should be detected in the output levels. An intermediate voltage level would cause the output driver to have a DC path through it and should be detected by an increased supply current as well.

Similar analysis can be carried out for the stuck-open faults on the wordline. It can be shown that a stuck-open fault on a wordline behaves like case (c) of the above analysis.

Multiple Access Faults: A multiple access fault occurs when more than the addressed cell are accessed during some cell operation. The decoder multiple access faults can be modeled as coupling faults in the matrix and need not be considered explicitly [33].

Coupling Faults: The algorithm covers all the modeled coupling faults. The performance of the 8N algorithm for modeled 2-coupling faults is shown in Table 5.2. The three columns of the table depict the coupling faults, when faults are sensitized, and when they are detected, respectively. For example, coupling fault 1, $(i = 0) \rightarrow (j = 0)$, is introduced by initialization step of march 1, as well as by that of march 2. However, it is not detected by march 1, since in this march, the coupled data (logic 0) is same as the original data in the coupled cell. This coupling is detected by march 2. Similarly, coupling fault 10 is sensitized several times but is detected only by march 4. The march 4 is added to detect this coupling fault $(i = x) \rightarrow (j = \bar{x})$. Owing to the nature of the

Table 5.2: Performance of the 8N DRAM algorithm for 2-coupling faults.

Coupling fault	Introduced in	Status
1	Init, march 1,2	Detected, march 2
2	Init, march 1	Detected, march 1
3	march 1,2	Detected, march 2
4	march 1	Detected, march 1
5	Init, march 1,2	Detected, march 2
6	Init	Detected, march 1
7	march 1	Detected, march 2
8	march 2,3	Detected, march 3
9	Init, march 1	Detected, march 1
10	Init, march 1,2,3,4	Detected, march 4
11	Init, march 1,2,3	Detected, march 3
12	Init	Detected, march 1
13	Init, march 1,2	Detected, march 2
14	Init, march 1	Detected, march 1
15	Init	Detected, march 1
16	march 1	Detected, march 1
17	Init, march 1	Detected, march 1
10	Init	Detected, march 1

coupling, this coupling fault can only be detected by a single element march in forward direction (address order $0 \to N - 1$). The complement of this coupling fault, coupling fault 12, is introduced by initialization step and is detected by the first march.

Data Retention Faults: The data retention for logic 1 can be covered by disabling the DRAM for the stipulated time and applying march 5. The data retention for logic 1 is tested by once again disabling the DRAM for the stipulated time and subsequently applying march 6. Alternatively, DRAM can be disabled after march 1 and 2, respectively, to test for data retention faults.

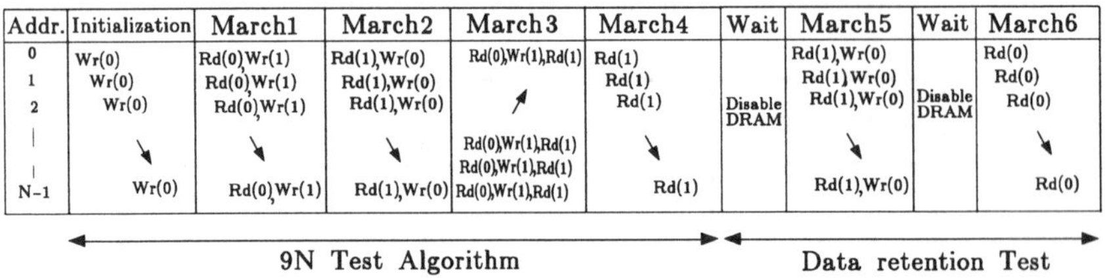

Addr.	Initialization	March1	March2	March3	March4	Wait	March5	Wait	March6
0	Wr(0)	Rd(0)Wr(1)	Rd(1),Wr(0)	Rd(0),Wr(1),Rd(1)	Rd(1)		Rd(1),Wr(0)		Rd(0)
1	Wr(0)	Rd(0)Wr(1)	Rd(1),Wr(0)		Rd(1)		Rd(1)Wr(0)		Rd(0)
2	Wr(0)	Rd(0)Wr(1)	Rd(1),Wr(0)	↗	Rd(1)	Disable DRAM	Rd(1),Wr(0)	Disable DRAM	Rd(0)
⋮	↘	↘	↘	Rd(0)Wr(1),Rd(1) Rd(0)Wr(1),Rd(1)	↘		↘		↘
N–1	Wr(0)	Rd(0),Wr(1)	Rd(1),Wr(0)	Rd(0),Wr(1),Rd(1)	Rd(1)		Rd(1),Wr(0)		Rd(0)

←——————— 9N Test Algorithm ———————→ ←——— Data retention Test ———→

Fig. 5.9: The 9N DRAM test algorithm with data retention test sequence.

5.4.5 Extensions

Dekker et al. [4] highlighted the problem of stuck-open detectability for sequential R/W logic. They suggested that adding one extra *read* operation in a 2-element march can ensure the transparency of the R/W logic for stuck-open faults. For such applications, march 3 of the algorithm is modified to include an extra *read* operation. The resultant algorithm, shown in Fig. 5.9, has the complexity of 9N. For word oriented DRAMs different data backgrounds are needed to cover intra-word faults [4]. If m bits per word are used, a minimum of K data backgrounds will be needed, where

$$K = \lceil \log m \rceil + 1$$

$$\lceil x \rceil = min\{n \varepsilon Z | n \geq x\}$$

(EQ 5.1)

When m is a power of 2, the formula simplifies to:

$$K = \log m + 1$$

(EQ 5.2)

However, different data backgrounds should be used only if bits constituting a word are adjacent to each other in the layout. On the other hand if different bits are not adjacent to each other, then the possibility of intra-word faults does not arise. Therefore, one data background is sufficient.

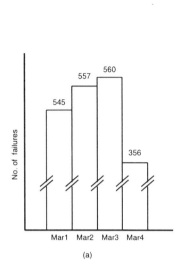

No.	Devices	Signature	Explanation
1	318	FFFF	Wordline stuck-at,read/ write failure,open
2	201	FFFP	Word, bit line failures read/write failures
3	21	PFFF	Bitline stuck-at 0, bit stuck-at 0
4	18	FPFP	Bitline stuck-at 1, bit stuck-at 1
5	11	PFPF	Coupling 3
6	3	PFPP	Coupling 1
7	3	FPFF	Combination of stuck-at 1 and coupling 10
8	3	FFPF	Combination of coupling faults 9 and 10
9	1	FFPP	Coupling 9

(a)   (b)

Fig. 5.10: (a) Number of failures per march element, (b) Different failure categories and explanations.

5.4.6 Results and Model Validation

The results obtained in previous sections are validated by applying the pro posed algorithm on Philips $4k \times 8$ embedded DRAM modules. The purpose of validation is to demonstrate the effectiveness of the algorithm and to verify the foregoing analysis. A large number of devices from 34 wafers is tested with the proposed 9N algorithm. Out of the total number of tested devices, 579 were failed by the algorithm. Fig. 5.10 shows the effectiveness of each march element in catching failed devices. For example, if a device failed in the first and third marches then the respective march failure numbers are increased by one. As expected, the most complex march element, march 3, caught the largest number of failures. In most cases, a large number of bits was failed which was also failed by the first or second march (Fig. 5.10(a)). Therefore, the performances of first and second marches in catching faults were comparable to that of march 3.

The table in Fig. 5.10 gives a concise analysis of failures. The first and second columns of the table show the type of failure and number of failed devices, respectively. In the third column, pass/fail response of the failed devices in terms of marches is shown. For example, FFFF means that the device is failed in all four marches. Similarly, FPFP means that the device is failed in the first and third marches and is passed in the second and fourth marches. The fourth column lists an explanation for each failure. As shown in this table, 318 devices failed in all marches. Bitmaps of the failed devices typically show wordline and/or bitline failures that could have been caused by bridging and open defects. Other possible causes include, read/write failures owing to the bridging of bit and $\overline{\text{bit}}$ lines, etc. Another set of 201 devices failed in first three marches, however, they passed the fourth march. In such cases, a typical bitmap showed wordline and/or bitline failures.

Twenty one devices were found to have a bit stuck-at-0 behavior. Therefore, except for the first march, these devices failed in the rest of the marches. Similarly, eighteen devices were found to be stuck-at-1. These devices passed second and fourth marches but failed in other two. These failures were caused by bridging or open defects in cells. An important difference between these two set of failures and failures mentioned previously was the number of failed bits. In the latter case, individual bits were failed, while in the former, majority or a large number of bits failed.

A small number of failures could not be explained by the fault model based upon catastrophic defects. However, these could be explained by 2-coupling faults model. For example, Eleven devices failed in second and fourth marches. This appears to be a stuck-at-0 behavior. However, if that is so, then devices should fail in the third march as well. This behavior could be explained by $(i=1) \rightarrow (j=0)$ coupling fault model (coupling fault 3). This coupling fault is initiated when a logic 1 in the coupling cell forces a logic 0 in the coupled cell. Since $i_{address} < j_{address}$ this coupling fault is not caught by the third march. As another example, a small set of three devices passed all marches except the second march. This is not a stuck-at-0 behavior. A stuck-at-0 behavior should also be detected by the third and fourth marches. However, this failure can be explained as a coupling fault 1. Three devices failed in all marches except in the second march. Their bitmaps revealed that in the first and third marches, a set of bits were failed and in the fourth march different set of bits were failed. Thus, these failures are thought to be a combination of the stuck-at-1 failures detected by first and third march elements and the coupling fault 10 detected by the fourth march element. Similarly, three more devices failed in all marches except the third march. However, in first two marches, one set of bits failed and in the fourth march a different set of bits failed. This behavior can be explained by a combination of coupling fault 9 and coupling fault 10. However, these coupling faults influenced different bits on the die. The coupling fault 9 caused a set of bits to fail in the first two marches but could not be detected by the rest of marches. The coupling fault 10 influenced another set of bits and could only be detected by march 4. Finally, one device failed only in the first two marches and its behavior coincided with that of coupling fault 9. In this fashion, all failures could be explained by the proposed fault model.

5.5 Address Decoder Faults

As pointed out earlier, the conventional wisdom suggests that RAM decoder defects can be mapped as RAM array faults that are detected during RAM array tests. Hence, no special test is needed for address decoders. However, we would like to reexamine this assumption as it is based on the analysis carried out in 70s [33,54]. Since then, semiconductor technology has changed significantly. Recently we came across some open defects in RAM address decoders that were not detected by linear test algorithms (e.g., march test) and resulted in field failures. This observation prompted us to look into occurrence of open defects in RAM address decoders.

Open defects or transistor stuck-open faults are known to cause sequential behavior in CMOS circuits and require 2-pattern test sequences for detection. The transistor and logic stuck-open testability has received considerable attention and a number of DfT solutions for stuck-open testing has been proposed (Chapter 2). However, application of these solutions to RAM decoders is not likely due to performance/area constraints. Furthermore, defects in address decoder are not directly observable. One has to excite them in such a way that they are detected via the *read* operation of the RAM. Finally, owing to the constraints of addressing sequence, the detection of open defects by march tests is not ensured. All these reasons put together, on one hand, render existing stuck-open DfT solutions for RAM decoders impractical, and on the other hand, make testing of such faults a new challenge.

A missing contact/via is a dominant source of open defects in CMOS technology. In the case of a DRAM process, the depth of contact is much higher compared to a logic process which increases the sensitivity for open defects. According to SIA technology roadmap for semiconductors, for a typical DRAM process the contact/via height/width aspect ratio is 4.5:1. The same for a typical logic process is 2.5:1 [45]. For future DRAM generations, the contact/via aspect ratio is expected to become 10.5:1, whereas for logic the same would become 6.2:1. The projected increase in the aspect ratio is a compromise to alleviate the large increase in per unit interconnect resistance and to prevent crosstalk [45]. Effectively, it means that for future CMOS devices in general, and DRAMs in particular, it will be much harder to make good, low resistance contacts. Furthermore, DRAM designs require tightest metal pitch and higher packing density, leaving no room for multiple contacts at most contact locations. Hence, open defects in RAMs require a careful investigation.

Open defects, in RAM matrix have been studied before and known to appear as cell(s), row/column read failures or cell(s) SAFs that may be detected by march tests. However, a class of open defects in address decoders are not detected by march tests. In this section, we focus on open defects in RAM address decoders and propose test and testability strategies for their detection.

5.5.1 Previous Work on Address Decoder Faults

A vast majority of the research on RAM testing was focused on efficient test algorithms for a variety of fault models. These fault models range from simple SAFs to complex pattern sensitive faults (PSFs) in the RAM array. However, little attention was paid to the faults in the address decoders or other RAM building blocks. Address decoder faults were assumed to be tested implicitly. An address decoder is a combinational circuit that selects a unique RAM cell for each given RAM address. Assuming that the faulty address decoder does not become sequential in operation, Thatte and Abraham [54] suggested that a faulty address decoder should behave in one of the following manners:

- The decoder does not access the addressed cell. In addition, it may access non-addressed cell(s).

- The decoder accesses multiple cells, including the addressed cell.

In the case of multiple accesses, the fault is viewed as RAM matrix coupling fault between different cells. In the case of no access, the cell is viewed as either SA0 or SA1. In simple terms, decoder faults manifest themselves as RAM matrix faults that are tested by the conventional algorithms.

5.5.2 Technological Differences

The above study was conducted for an NMOS decoder. The open defects in NMOS address decoders cause a logic SA behavior. As the technology made transition from NMOS to CMOS, the validity of the assumption was never re-evaluated. In CMOS technology, only a subset of open defects cause a logic SA behavior. The rest of the open defects cause sequential behavior in logic gates. Some of the defects causing sequential behavior in address decoders may escape detection by conventional tests.

The difference between an NMOS and a CMOS address decoder can be explained with the help of Fig. 5.11. The figure shows a typical address decoder and logic implementations in NMOS and CMOS technologies. A logic gate in NMOS technology is implemented by a depletion mode NMOS load transistor and switching enhancement mode transistors. On the contrary, a logic gate in fully static CMOS technology is implemented by equal numbers of enhancement mode PMOS and NMOS transistors.

An address decoder selects a specific wordline depending on the given input address. This requires the output of a logic gate in the address decoder to be

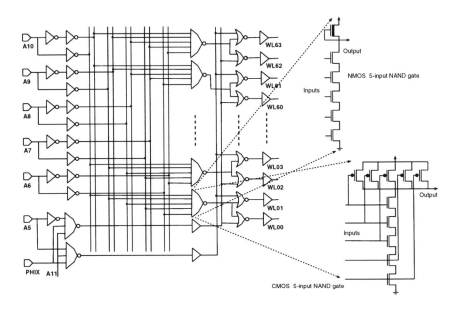

Fig. 5.11: A typical RAM address decoder with logic implentations in NMOS and CMOS technologies.

active only for a unique input address and remain inactive for all other addresses. For example, for the 5-input NAND gates in Fig. 5.11 the output is active (logic 0) only if all inputs of the gate are high and the output is inactive (logic 1) for the rest of the cases. In the case of NMOS technology, the depletion mode load transistor pulls up the output to the inactive state when inputs are not causing the gate to be in the active state. Now, an open defect in a switching transistor of the NMOS logic gate will cause the gate to stay in the inactive state when it was suppose to be in the active state. In other words, such a defect will cause the addressed decoder not to access the addressed cell. On the other hand, if there is an open defect in the load transistor, the logic gate will stay in the active state causing a multiple access fault.

In the case of the CMOS address decoder, the active state is arrived in the same manner. However, the inactive state is reached by several parallel paths (depending upon the fan-in) selected by the input addresses. In later sections

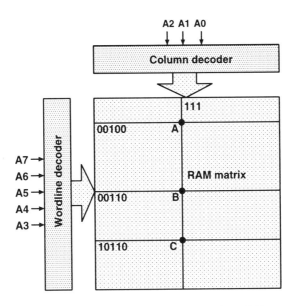

Fig. 5.12: A graphical representation of the failure.

we shall discuss the fault behavior that open defects in these parallel inactive paths can cause.

5.5.3 Failure and Analysis

Undetected faults of address decoders are explained in Fig. 5.12. The figure depicts a part of an embedded SRAM block diagram showing the matrix, wordline and column decoders. An actual failure mechanism in an embedded SRAM is illustrated with the help of three cells, A, B, and C, respectively. The addresses, (A7--A0), of these cells are 00100 111; 00110 111; and 10110 111, respectively. The considered SRAM has 256 addresses (8-bits) and the word is also 8-bit wide. Different bits are not close to each other, so there is no possibility of an intra-word *coupling fault*. Address bits A7, A6, A5, A4, and A3 decode the word lines and the rest of the bits select the column (or bit) line. Cell C is the cell that fails conditionally. Following are the symptoms of the failure as observed:

1. Write address C (10110 111) with logic 1.

2. Write address A (00100 111) with logic 0.

3. Read address C: result is logic 1, that is correct.

Observation: RAM behaves normally, because address input A4 has changed.

4. Write address B (00110 111) with logic 0.

5. Read address C: result is logic 0, that is wrong!!

Observation: the failure occurs, because no address input among A3, A4 or A5 has changed.

The *read* operation on cell C yields wrong data value only if between the *write* and *read* operations for cell C, some address bits (A5, A4 and A3) are kept unchanged. If any of these bits are changed, then the *read* operation for cell C yields the expected data value. Furthermore, the fault is completely data independent. The failure does not *write* data into another cell and appears as a *read only* error.

5.5.4 Analysis

From the above mentioned failure symptoms following deductions are made:

1. All three cells have the same column address (111).

2. Cell C yields a *read* failure when address bits A5, A4 and A3 are kept unchanged.

3. The fault causes only a *read* failure in cell C and does not influence other cells in any manner.

4. The fault is not detected by the 6N SRAM test algorithm.

The first deduction suggests that when cell B is enabled after a cell C access, somehow cell C is also enabled (or it is not disabled). Cell C is controlled by the corresponding wordline. Consider a situation when wordlines B as well as C are enabled. If the complementary data (cell C) is written in cell B, the same is written in cell C as well. Hence, a subsequent *read* operation on cell C results in a *read* failure. The second deduction makes it clear that it is not the case with all cells of the same column. Cell C is sensitive only when address bits A5, A4 and A3 are not changed. The third deduction strengthens the first one stating that only a *read* on cell C is affected by the defect mechanism.

There are two possible explanations that match the above symptoms and deductions: (i) the wordline of cell C is also enabled when wordline of cell B is enabled, and (ii) the wordline of cell C is not disabled when cell B is enabled. The first possibility is unlikely. However, it could be caused by (a) a decoder design error or (b) a low resistance bridging fault between wordlines of cells B and C. The decoder design error is ruled out since in that case a large number of devices would then fail under the test conditions. The low resistance bridging fault explanation also seems unlikely since, the corresponding fault should be bidirectional. Moreover, such a defect is a typical case of a decoder fault mapped onto the matrix coupling fault that should be detected by the 6N test algorithm. Therefore, the fault that wordline of cell C is not disabled when cell B is selected is a likely cause.

The argument that cell C is not disabled can be explained with the help of Fig. 5.13 that illustrates a part of the wordline address decoding logic and the corresponding bitlines. Fig. 5.13 does not show the wordline drivers and input buffers. The wordline decoder has a 5-bit address. The address bits are buffered and their true and complement values are generated. The address decoding is achieved with the help of 4-input NAND gates. Subsequently, 3-input NOR gates decode the outputs of NAND gates with address bit A6. A periodic, timing signal, PHIX, forms the third input to these NOR gates. The outputs of NOR gates are buffered to drive the wordlines.

Let us assume for a moment that the NAND gate in the wordline decoder that decodes cell C has an open defect such that a p-channel transistor having A7 as its input is disconnected from VDD. Now, let us once again try to repeat the experiment carried out earlier. It is easy to notice that all the steps and observations can be re-created with the defect in cell C. In a decoder constructed with NAND gates, a simultaneous logic high on all inputs cause n-channel transistors to be in conduction mode and hence enables the particular wordline. For example, 1111 on A7, A5, A4, $\overline{A3}$ and 0 on A6 select the wordline corresponding to cell C. However, a disabling of that particular path can take place by four (or depending upon the fan-in of the NAND gate) paths. Suppos-

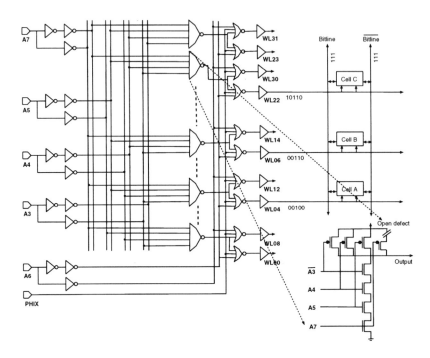

Fig. 5.13: An undetected open defect in the address decoder.

ing if one of the paths has an open defect, the wordline can not be pulled high (disabled) through that path. If the wordline is disabled through the faulty path (for example, by selecting cell B), two cells are selected at the same time. Therefore, a *write* operation on another selected cell is also performed. Now, if a *read* is performed for cell C, depending on the original stored data value and new data value, a fault is detected. However, if cell A is accessed after cell C, a parallel p-channel path in the faulty NAND gate will disable the corresponding wordline and the fault will not be activated.

The subsequent analysis demonstrated a transistor stuck-open fault caused by missing source to VDD contact. Such a fault can be caused by several defects. A missing contact between source (drain) diffusion and Metal1 is the most likely cause. An open defect in the metalization layer (step coverage problem) causing a transistor stuck-open fault is another possibility.

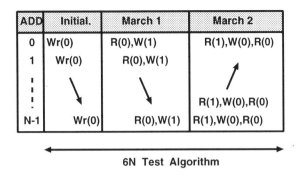

6N Test Algorithm

Fig. 5.14: The 6N SRAM march test algorithm.

5.5.5 Why Non-detection by March Tests?

The question is why the fault is not detected by the 6N march test and how should we detect such failures. The 6N test algorithm is shown in Fig. 5.14. It is a popular and time tested algorithm used within Philips for SRAM testing [29]. First, the RAM is initialized with logic 0. Subsequently, March1 reads the initialized value and writes logic 1 in each RAM cell in ascending address order. The following binary address after wordline C (address 10110) is 10111, which modifies the A3 bit. Hence, wordline C is disabled like a fault-free case (Fig. 5.13). In other words, the fault is neither activated nor detected in March1. Similarly, the fault is not detected in March2. The March2 is in descending address order. After the wordline C is activated, the next wordline address (descending order, 10101), which modifies A4 and A3 bits. As a result wordline C is disabled once again and the fault is not detected. This type of fault can only be detected by a march test (or a linear algorithm), if the next wordline address causes the fault to be activated in at least one march direction, and keep it activated till it is detected by a *read* operation on the cell. Now, depending upon the original and over-written data values, the defect can be detected. However, this condition can not be met for all such open defects in NAND gates. Therefore, most of such defects are not detected by march tests.

A march test may have any address order, as long as all addresses are access-

ed. For reasons of simplicity, mostly ascending (descending) address is selected. However, without the loss of generality, it can be argued that no addressing
sequence will detect all open defects in the address decoder. Furthermore, no
linear test algorithm will detect such defects since due to these defects, the address decoder is changed into a faulty sequential circuit that, in general, requires a two-pattern test. The basic assumption about address decoders that
under faulty conditions they should remain combinational is violated. This is a
generic problem with decoders implemented with static CMOS logic gates.
When decoders are implemented with dynamic logic (or NMOS) such faulty
conditions may not arise.

5.5.6 Address Decoder Open Defects

A linear, march test algorithm will not detect some open defects in address decoders. Other RAM test algorithms are also not likely to detect these defects
owing to the fact that two-pattern test sequence (T1, T2) for all potential defects is not ensured by them. For example, complex algorithms for neighborhood pattern sensitive faults will not be able to ensure decoder open fault
detection. Arguably, GALPAT (GALloping PATtern) algorithm [11] of complexity $O(n^2)$ will detect these defects. However, application of GALPAT even
for moderately sized RAMs is not possible due to its excessive test time. On
the contrary, we shall see later in this chapter that there has been a significant
effort made to employ parallel test techniques to reduce RAM matrix test
costs. In parallel test techniques, address decoder faults are less vigorously
tested. As a result, RAM address decoder testing becomes a quality and economics issue.

5.5.6.1 Open Defects in an Address Decoder

Only a subset of all open defects in an address decoder is not detected by
march tests. Hence, it is logical to analyze which open defects are not likely to
be detected by march tests and then devise a test only for those open defects.
For this purpose, we take a row decoder of an embedded RAM (Fig. 5.15).
This circuit decodes a 6-bit address (A10 -- A5) to 64 wordlines. Address bit
A11 determines the selected quadrant and PHIX is a periodic timing signal

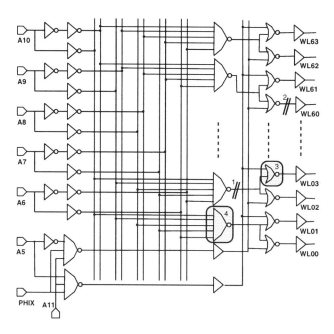

Fig. 5.15: A typical wordline address decoder.

that controls the timing of the wordline (X) address. In this decoder, instead of 4-input NAND gates 32 5-input NAND gates are utilized and the 6th bit (A5) is further decoded by 64 2-input NOR gates.

In general, open defects in an address decoder may occur either in between logic gates (inter-gate) or inside a logic gate (intra-gate). Defects 1 and 2 in Fig. 5.15 are representatives of the inter-gate class. The inter-gate open defects cause a break in an interconnect line. Owing to this class of defects, at least one RAM cell cannot be addressed. Hence, a cell may appear to have a stuck-at fault. In other words, inter-gate open defects do not cause sequential behavior and they are detected by march tests such as the 6N algorithm. However, intra-gate open defects (defects 3 and 4) are difficult to detect because they may influence only a single transistor. Hence, they may result in a sequential behavior. If an intra-gate open defect disconnects all paths between the output and VDD (VSS), it effectively causes an output SA0 (SA1) fault that is detected by the march test. However, if an open defect disconnects only one of the paths between the output and VDD (VSS), it causes a sequential behavior.

Let us assume that defect 3 causes an n-channel transistor in a 2-input NOR gate to be disconnected from VSS. Since, there is only one other n-channel transistor in parallel with the defective transistor, the defect will be detected by either the ascending march (March1) or the descending march (March2) of the address space depending on which of the transistors is faulty. The condition for this detection is that the inputs of the faulty gate should be changed in a Gray code manner. Therefore, in this case address bit A11 and decoded A10--A5 bits should change in the Gray code manner. The situation becomes complex as the number of inputs in a gate rises to three or more. With the reasoning of the previous section, it can be concluded that detection of all open defects in a 3 (or more) input logic gate is not guaranteed by the march test. It can be observed in this example, that at least 3 open defects in each of the 5-input NAND gates will not be detected (the other 2 open defects will be detected by descending or ascending march elements). There are 32 such NAND gates in the decoder giving rise to at least 96 undetected defects.

5.5.6.2 Supplementary Test Algorithm

Once all likely escapes are known, a test solution may be devised. A small algorithmic loop is appended to the 6N algorithm to detect address decoder stuck-open faults not detected by the 6N algorithm. However, this algorithmic loop is specific to an address decoder and is independent of the 6N algorithm. Hence, it can be added to any other test algorithm.

Let us assume that M is the number of input bits of the wordline decoder and the number of wordlines equals 2^M. To test the row decoding logic we can select any arbitrary column address for *read* and *write* operations. In the following algorithm we set the column address to 0. As explained before, the least significant bit (in Fig. 5.15, bit A5) is a don't care and remains 0 during the test. To test for the hard-to-detect opens, the NAND gates in the decoding logic should be tested in a sequential manner. For each NAND gate a logic 0 is written in the selected cell (say D) by the corresponding wordline (remember that bit A5 is set to 0). Subsequently, the wordline address is changed such that only one address bit is changed (let us say A6). This will allow the particular NAND gate to be disabled through one selected p-channel transistor. Now, logic 1 is written in the new address location (say E). If the selected p-channel transistor had an open defect, the cell D is still enabled and the *write* operation on cell E can also over-write the content in cell D. A subsequent *read* operation on cell D will detect a *read* failure and hence the open defect. This is repeated for all address bits to NAND gates and for all NAND gates. For example, for the 5-input NAND gate in Fig. 5.15 with defect 4 (shaded) following test sequence can be applied:

1a. Keep Y decoder address constant,
1b. keep A5=0 and A11 (if available)=0

2a. Let A10A9A8A7A6 = 00000, Write(1);
2b. A10A9A8A7A6 = 00001, Write(0);
2c. A10A9A8A7A6 = 00000, Read(1);
2d. A10A9A8A7A6 = 00010, Write(0);
2e. A10A9A8A7A6 = 00000, Read(1);
2f. A10A9A8A7A6 = 00100, Write(0);
2g. A10A9A8A7A6 = 00000, Read(1);
2h. A10A9A8A7A6 = 01000, Write(0);
2i. A10A9A8A7A6 = 00000, Read(1);
2j. A10A9A8A7A6 = 10000, Write(0),
2k. A10A9A8A7A6 = 00000, Read(1);

In general, an algorithm for a given address decoder can be evolved that will supplement any RAM test algorithm. For the address decoder in Fig. 5.15, such an algorithm is shown below:

In the algorithm description the address values in the *read* and *write* operations correspond to the binary code at the input bits of the wordline decoder (A10, A9, A8, A7, A6, A5). The algorithm becomes:

Column_address = 0
For** i = 0 **to** 2^{(M-1)} **Do
    *Base_address = 2 * i*
    Write "0" to Base_address
    For** j = 0 **to** M **Do

$Write\_address = Base\_address\ XOR_{binary}\ 2^j$

$Write\ "1"\ to\ Write\_address$

$Read\ "0"\ from\ Base\_address$

   End For

End For

As can be determined from the algorithm, the inner loop will be executed M-1 times and for each i it consists of one *write* and one *read* operation. The main loop will be executed $2^{(M-1)}$ times and takes one extra *write* operation. This makes a total complexity of the algorithm as $(2M-1) \times 2^{(M-1)}$ *read* or *write* operations, where M is the number of input bits in the wordline decoder. To compare the complexity of the algorithm given above with the 6N algorithm we will consider a RAM having 6 bits devoted to the column decoding and another 6 bits to wordline decoding. The 6N algorithm will take $6 \times 2^{12} = 24,576$ *read* and *write* operations. The algorithm given above will only take $11 \times 2^5 = 352$ *read* and *write* operations. So the additional test complexity is less than 2% of the 6N test.

It can be argued that similar open defects in the column decoder can also cause hard to detect faults. Column decoders can be analyzed to devise a suitable test algorithm.

5.5.7 Testability Techniques for Decoder Open Defects

In the previous section we discussed the test escape problem arising due to open defects in RAM address decoders and evolved a test procedure to prevent test escapes. In this section, we focus on the layout level testability measures to simplify the detection of such hard to detect open defects and on building fault tolerance through logic modification. Such measures are best implemented while designing new address decoders. If these simple yet effective measures are implemented, the requirement for additional test is either completely eliminated or drastically reduced. However, the existing decoders without such measures will require the extra test procedure as proposed previously.

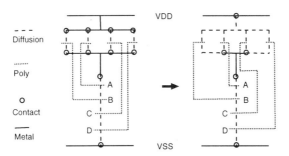

Fig.5.16: Layout transformation of a 4-input NAND gate for open
defect testability.

5.5.7.1 Layout measures

The layout improvement is probably the simplest and most effective method
to reduce the occurrence of such open defects. The layout of the circuit affects
the testability to a great extent. Simple layout modifications may reduce the
possibility of hard-to-detect faults, hence, reducing the burden on test generation. For example, Placement of multiple contacts at hard-to-detect defect locations (parallel transistors) in the decoder will make it robust against open
defects. These layout techniques are well documented in the literature [19,22].
There is a need to implement such techniques in future RAM decoder designs
because (i) the decoder circuitry is implicitly tested by testing only the matrix,
(ii) the RAMs are often tested by linear algorithms that restrict the excitation
of the decoder in a particular fashion so as to cover the fault model in small
number of operations, and (iii) there are quality and economic issues.

The layout measures are explained with the help of Fig. 5.16. The figure illustrates a switch graph representation of a 4-input NAND gate. This transformation is similar to the one proposed by Levitt and Abraham [22]. In the
unmodified layout an open at a contact can occur at any branch of parallel
transistors or metal lines. A simple test may not detect such an open defect,
and all the parallel branches must be tested separately. We assume that the
open defect probability due to a poor contact is relatively high compared to
open defect probability due to a break in diffusion. The modified layout re-

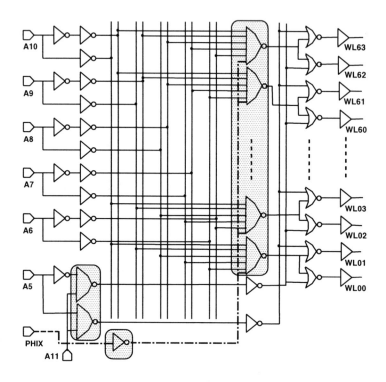

Fig.5.17: A fault tolerant row decoder against hard to detect open defects.

sults in a robust design as well as in simpler test generation for open defects. It was reported [22] that though area and delay of the transformed gate may increase marginally, the number of hard to detect faults is reduced drastically.

5.5.7.2 Logical measures: building fault tolerance

The layout level techniques, in principle, can reduce the probability of occurrence of open defects in sensitive decoder locations but cannot eliminate them completely. Therefore, in this subsection we propose methods for implementing fault tolerance in the key decoder locations such that in spite of the defect, the decoder and the RAM can function correctly. The fault tolerance can be used together with the layout level transformations to enhance the robustness of the decoder. Fig. 5.17 illustrates the concept of logical measures. This figure is the same as Fig. 5.15 except for the highlighted areas, an inverter and an added net shown by a broken bold line. From the earlier reasoning, we conclude that opens affecting only single p-channel transistors in 5-input NAND gates are hard-to-detect. The p-channel networks in NAND gates provide the disabling paths to the wordlines (since a particular wordline is selected if and only if the output of the corresponding NAND gate is logic 0). Therefore, an extra p-channel transistor can be added in each of the 5-input NAND gates such that it provides an alternative path for wordline disabling such that before application of a new address all NAND gates are disabled. In other words, no wordline is selected. A corresponding n-channel transistor is also added to avoid logic conflicts, effectively making it a 6-input NAND gate. The modified NAND gates are shaded in the figure. The inputs of these transistors are driven by the PHIX signal which activates the wordline address. Effectively, PHIX now gates address bits A10----A6 instead of A5 and A11 (see Fig. 5.14). The extra inverter is needed to invert timing signal PHIX. In a decoder where 5-input NOR gates are utilized instead of NAND gates, the extra inverter is not needed. As far as the logic function of the decoder is concerned, it is not changed. The design of the decoder can be optimized for correct timing without sacrificing the gains. Furthermore, highlighted 3-input NOR gates are reduced to 2-input NOR gates

5.6 Parallel Testing of RAMs

In recent years, owing to the increasing capacity of each successive RAM generation, the reduction of test time (cost) has become an important issue. A recent study of DRAMs identified the test cost, along with process complexity, die size and equipment cost, as a major factor contributing to the future DRAM chip cost [17].

The access time of DRAM decreases 0.8 times for every new generation of DRAM while the number of bits increases by a factor of four. Effectively, the testing time increases by a factor of 3.2 for each new generation. This results in tremendous increase in test cost that prevents the cost per bit from coming

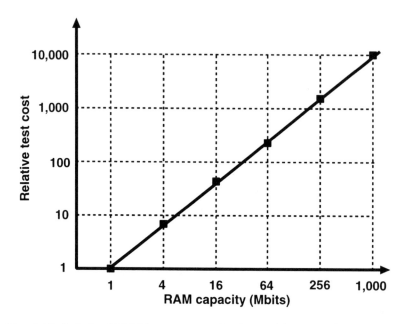

Fig. 5.18: Relative DRAM test cost projections without parallel testing [38].

down in spite of the increased integration. The test cost of a 64 Mbit DRAM is projected to be 240 (see Fig. 5.18) times that of 1 Mbit DRAM. For a 64 Mbit DRAM, the test cost ratio of total product cost is expected to be 39% (see Fig. 5.19). Furthermore, if conventional test methods are used, test cost will grow at a rapid rate [17]. The SRAM test cost is also expected to follow a similar trend. Therefore, testing of RAMs in an efficient, reliable and cost effective manner is becoming an increasingly challenging task.

Parallel testing of RAMs has the potential of reducing its growing test cost. RAMs have an architecture that is well suited for parallel testing. Instead of testing one bit at a time, a large number of bits can be tested together. As a consequence, several attempts have been made to adopt multibit parallel test-ing of RAMs [16,17,20,25,26,28,38,43,47,50]. Kumanoya et al. [20] were probably the first to apply the parallel test concept to a 1 Mbit DRAM. They reduced the test time to a quarter by testing four bits simultaneously and named their method as multi-bit test (MBT). In this approach, four partitions

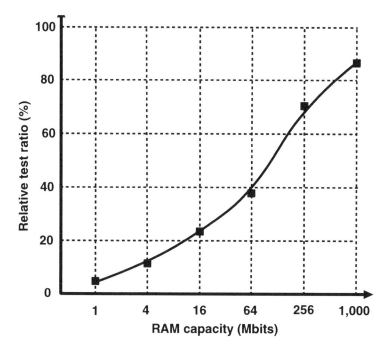

Fig. 5.19: Relative DRAM test cost (%) projections without parallel testing [38].

were used in a 1 Mbit DRAM to reduce the test time by a factor of four. The MBT mode became popular and was used in 4 Mbit DRAMs as a standard function. Similarly, to reduce overall test time and cost, many RAM chips can be tested in parallel, sharing resources of a single test system [38].

Sridhar [47] proposed the usage of on-chip logic to reduce the test time both for a given RAM test algorithm and for a given chip. This approach relies on parallel signature analyzers (PSAs) or a multiple input shift register (MISR). A PSA is a multiple-input linear feedback shift register (LFSR) used to compress test response data in a digital circuit under test. McAdams et al. [28] designed a 1 Mbit DRAM with a similar DfT scheme. The RAM matrix was partitioned into eight sub-arrays that were tested concurrently. In their scheme, data was written simultaneously on eight cells, identically located inside sub-arrays. During the read operation, when any of the eight cells is addressed, all of them are simultaneously accessed and their contents are compared by a two-mode 8-bit parallel comparator. This method reduces the test time by a factor of 5.2. Shah et al. [43] extended the concept to 4 Mbit DRAMs with 16 bit parallel comparator. The test complexity of the 4 Mbit DRAM was reduced to that of a 256 Kbit DRAM.

The MBT mode is realized by using nibble mode circuits. The number of bits tested in parallel at one time is limited to 16 even in 16 Mbit DRAMs in view of the additional circuit and the memory architecture. Inoue et al. [16] have pointed out that in the case of the MBT mode it is not possible to test a large number of bits in parallel without incurring a huge area penalty. Therefore, such a scheme for 16 Mbit or larger RAMs will not be appropriate. To reduce the test time drastically, they proposed wordline test mode. They called it the *line mode test* (LMT). An LMT tests all bits connected to a wordline simultaneously. The on-chip test circuit provides parallel write and parallel compare possibilities. The write circuit is used to write a given data to a wordline. The parallel compare circuits are used to compare data from bits with expected data and to generate a flag F indicating whether or not a fault has occurred. In the LMT, all bits are tested in cycles equal to the number of wordlines. Therefore, pattern length of this test is proportional to $N^{1/2}$.

The line mode test has its shortcomings. The application of an arbitrary parallel data pattern on the complete wordline is not possible due to the constraint of the parallel write circuit. Therefore, the test mode cannot adequately test coupling or pattern sensitivity faults in RAMs. Some of these constraints were removed by Matsuda et al. [25]. They applied the LMT principles to a 16 Mbit DRAM using a new array architecture combined with an on-chip test circuit suitable for generating random pattern tests for the wordline. High fault coverage is achieved through the application of random patterns. Such a test scheme became popular for 16 Mbit DRAMs supplied by many Japanese companies. As mentioned previously, DRAMs are very sensitive to parametric or soft faults. Mazumder [26] proposed a DfT scheme based on LMT principle for parametric faults.

The implementation of any parallel RAM test scheme requires a careful consideration of several issues. High speed access as well as low power consumption must be maintained during the parallel testing. Moreover, test time reduction should be significant, performance impact should be minimal, and area overhead of the scheme should be acceptable, in order to implement these schemes. Considering these constraints, Takeda et al. [50] outlined a parallel test scheme for 64 Mbit DRAMs. They argued that the conventional type of parallel processing test suffers from two problems at the production level. First, the access time is delayed due to a circuit that checks if all the multiple bits are active together. Second, the multiplicity of bits is insufficient for a 64 Mbit DRAM. If the conventional techniques of multiplicity are used for larger number of bits, the width of the bus increases resulting in unacceptable area increase and increased test mode power dissipation. To counter this, if the power dissipation is lowered, the access time is increased. They suggested that 64 Mbit DRAM test complexity can be divided into two main parts: (i) testing of peripheral circuits, and (ii) testing of DRAM cells. The former checks for the timing of each cell's operation, for which a common high speed access is required. The later test is concerned with the data retention characteristics of DRAM cells, using a high multiplicity. They implemented a DfT technique that has high speed compression and highly multiplexed compression capabilities.

Instead of putting test hardware on a RAM, Inoue et al. [17] suggested fabrication of a test acceleration chip (TAC) for a low cost memory test. A matrix of such chips is constructed and acts as the interface between the tester and the matrix of DUT array. Through this concept 100 or more RAM chips could be tested in parallel. Each TAC has a built-in digital timing comparator that tests for AC parameters of the target RAM device. Though, the test time for each chip remains the same, the parallel testing of chips reduces the test costs significantly. For a 100 TAC array, the cost is reduced by a factor of 100. They envisaged that such a DfT scheme will be suitable for 64 Mbit, 256 Mbit and 1 Gbit DRAMs.

5.7 I$_{DDQ}$ Based RAM Parallel Testing

The increasing test costs of RAMs necessitates a fresh approach towards testing. RAMs are the densest and one of the largest chips ever to be tested. For example, a DRAM chip size has grown by 40 to 50% while the cell size has decreased by 60 to 65% for each successive generation [17]. The chip size of 64-Mbit DRAM is expected to be in the range of 200 mm^2. Small feature size and huge chip size result in an enormous critical area [7] for defects. Defect-free yield of big RAMs is low. Therefore, redundancy techniques are utilized to achieve an acceptable yield. In addition, RAMs are mass produced. The conventional voltage based March tests when applied to Mbit RAMs are prohibitively expensive. Therefore, RAM test strategies are under severe pressure to ensure the quality of the tested devices while maintaining the economics of the production.

RAMs have a well defined architecture that is suitable for I$_{DDQ}$ based parallel testing. Relatively large number of bits can be tested in parallel without extensive circuit modifications. I$_{DDQ}$ testing is ideally suited for parallel testing because the faulty (or fault-free) information traverses in parallel through the power buses. Therefore, no special observability conditions are needed. This property of I$_{DDQ}$ testing has been extensively utilized to reduce the test costs of digital VLSI. Nevertheless, a straightforward application of I$_{DDQ}$ test technique to RAMs has a rather limited defect detection capability. Due to the very nature of RAM architecture and its operation, the I$_{DDQ}$ test is not able to detect many of the otherwise I$_{DDQ}$ detectable manufacturing defects. In other words, the full potential of I$_{DDQ}$ testing for parallel RAM testing is not exploited.

The I$_{DDQ}$ test coverage of manufacturing process defects is enhanced extensively with minor RAM design modifications. The RAM address decoders, bitline precharging circuitry and control unit are modified so as to create an I$_{DDQ}$ test mode. In this test mode a majority (or all) bits can be tested in parallel. Therefore, number of required I$_{DDQ}$ measurements is reduced drastically, making the I$_{DDQ}$ based RAM production testing practical. A vast majority of manufacturing process defects is tested with few I_{DDQ} measurements. Those defects that are not detectable by the I_{DDQ} technique are detected by voltage based march test. The combined application of I_{DDQ} and march tests reduces the RAM test complexity compared to any published RAM test algorithm while improving the test quality considerably.

5.7.1 Review of RAM I_{DDQ} Testing

There have been attempts to use I_{DDQ} test method for detecting manufacturing process defects in SRAMs [12,15,29,31,49,56]. Horning et al. [15] found that after the dynamic voltage stressing, I_{DDQ} testing caught approximately 38% of the failed devices over and above those caught by the functional test. Although the reported results show the effectiveness of the I_{DDQ} test method, there are manufacturing process defects in an SRAM cell matrix that are not detected by the I_{DDQ} test. Meershoek et al. [29] found many devices that passed the I_{DDQ} test, but failed functionally. Su and Makki [49] utilized dynamic power supply current to detect pattern sensitive faults in a SRAM matrix. They employed a novel power grid structure so that switching current transients are isolated from the dynamic current resulting from pattern sensitive faults. Thus, switching current transients would not invalidate the test.

Naik et al. [31] carried out a detailed failure analysis of a CMOS SRAM. They reported that some bridging defects in and around a SRAM cell did not give rise to the high quiescent current. Therefore, these defects were not detected by the I_{DDQ} test. However, the dynamic *write* cycle (W0 or W1) current was high and gave an indication of defects. In general, RAMs have large bitline capacitances. Therefore, the dynamic current is not a reliable and robust measure for defect detection. In addition, it becomes less accurate as the speed of RAM operation is increased.

Yokoyama et al. [56] recently presented a concept similar to the one presented by us in this chapter. They analyzed the behavior of a single SRAM cell in the presence of various defects and found that most of them (76%) affect power supply current when a *write* operation is executed. Based upon this observation, they suggested modifications of address decoders so that whole SRAM matrix is treated as a single SRAM cell in the test mode. Then, if a *write* operation is performed, many defects will be detected by an elevated I_{DDQ}.

It can be argued that this approach will detect only those intra-cell defects that give rise to an elevated I_{DDQ}. Such defects include, shorts causing cell SAF and gate oxide defects causing cell leakage faults. On the other hand, an inter-cell bridging defect causing a coupling fault will not give rise to an elevated I_{DDQ}. In order to increase the defect coverage of the I_{DDQ} test, the address decoders need to have extra functions such that inter-cell defects can also be detected. Sachdev [41] presented one such DfT scheme. He took the realistic SRAM fault model developed by Dekker et al. [4] and proposed an I_{DDQ} based test methodology for the fault model. Those defects/faults that are not detectable by the I_{DDQ} test technique are detected by the conventional march test. He gave a comparative analysis showing the RAM test complexity reduction and addressed practical aspects of the implementation.

5.7.2 I_{DDQ} Testing and RAM Defects

The I_{DDQ} detection of a bridging defect depends on the ability to control different signals to desired logic levels so as to create and sustain a logic conflict across the defect. This logic conflict gives rise to the quiescent current which may be detected by an I_{DDQ} measurement. The logic conflict has to be sustained for long enough time to allow an I_{DDQ} measurement. A bridging defect across which a logic conflict cannot be created and sustained due to architectural or functional constraints, is not detected by I_{DDQ} testing. For example, a bridging defect across any two nodes A and B is not detectable with I_{DDQ} measurement if at any given moment only one node is driven to a logic value and other node is in high impedance state.

A similar situation occurs in RAMs. Let us assume that a DRAM cell C has BL_i and WL_j as its bitline and wordline, respectively. While performing a *write* operation on C, only BL_i is driven from input and the remaining bitlines are either in a high impedance state or performing a refresh operation on other cells on WL_j. A bridging defect between BL_i and BL_{i+1} does not cause a sustained logic conflict. Hence, the defect is not detected by an I_{DDQ} measurement. Control timings of RAMs are internally generated by a control circuit. Often, this control circuit is implemented as an asynchronous delay line, consisting of delay elements and set-reset flip-flops. This is done to optimize the power consumption and the speed of operation. Therefore, internal timing can not be controlled externally. The control of internal timings is crucial for effective I_{DDQ} measurements. For example in DRAMs, during a *write* operation on a cell, the corresponding pass transistor is enabled only for a short time (few tens of nanoseconds, typically). During that time the cell is being driven. If a DRAM cell contains a leakage defect, it will cause a leakage only for that same duration. This length of time is not sufficient for a reliable and unambiguous I_{DDQ} measurement.

A similar explanation also holds for SRAMs. Most of the SRAMs are asynchronous that means they do not require external clocks. On the other hand, synchronous SRAMs do require external clock. The advantages of synchronous SRAMs are the speed of operation, latched input/output data, etc. [37]. Nevertheless, the problem of poor defect detection remains, irrespective of the SRAM architecture. Even though it is possible to control the internal timing of a RAM, it is impractical to perform an I_{DDQ} test on each address. Also, with successive RAM generation it will become increasingly expensive. Therefore, a low cost DfT strategy is needed for I_{DDQ} testing of RAMs.

5.7.2.1 SRAM Defects

Fig. 5.20(a) illustrates a part of an SRAM block diagram. The cells are organized in rows and columns, and are connected to their respective bit (BL), $\overline{bit}$ (BLB), and wordlines. The precharge circuit raises BL and BLB lines to a predefined voltage. The sense amplifier elevates the cell logic value to full CMOS levels. During a *read* operation on a cell, following actions are carried out sequentially:

1. All bitlines are precharged to a specified voltage.

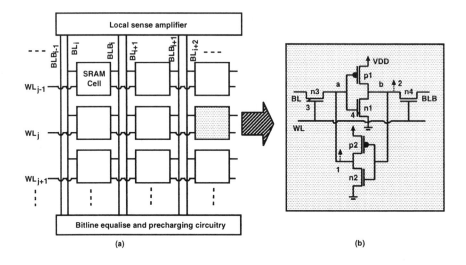

(a)   (b)

Fig. 5.20: (a) A part of SRAM block diagram, and (b) an SRAM cell and some I_{DDQ} undetectable defects.

2. The row decoder enables the appropriate wordline allowing it to read data from the cell.

3. The column decoder selects the appropriate BL and BLB lines to be connected to the sense amplifier.

4. The sense amplifier amplifies the signal to CMOS levels

A *write* operation is similar in sequence. We have seen in the previous chapter that some low resistance bridging defects in CMOS flip-flops, causing stuck-at faults, are not detected by I_{DDQ} measurements [21,40]. Latches in the flip-flops are similar to an SRAM cell. Following a similar reasoning, it can be concluded that SRAM cells have some bridging defects not detected by I_{DDQ} testing. For example, a low resistive bridging defect, defect 1 in Fig. 5.20(b), between node 'a' and VDD is not detected by I_{DDQ} testing. This defect causes the cell to have a SA1 behavior. In order to detect it, the corresponding bitline should be driven to logic low and the corresponding wordline should be activated for the duration of the I_{DDQ} measurement. Since internal timing signals are generated asynchronously, the wordline is activated only for a short time,

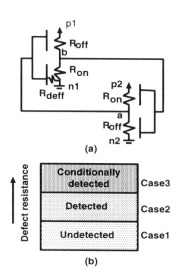

(a)

(b)

Fig.5.21: (a) Equivalent circuit for an SRAM cell, and (b) Gate oxide defect detection depending upon its resistance.

typically for few tens of nanoseconds. The quiescent current measurement often requires an elapsed time that is an order of magnitude larger then few tens of nanoseconds. Thus, I_{DDQ} test can not detect the defect. The defect would have been detected, if the timing of the SRAM was controlled externally. Besides the control of the internal timing, the I_{DDQ} measurement is needed on each SRAM address. This is an impractical solution for high density SRAMs. Similarly, defect 2 is not detected by I_{DDQ} testing. A gate oxide defect on transistor 'n3' (defect 3) remains undetected because the voltage across the defect is not sustained for a long duration. The case of a gate oxide short in transistor 'n1' (defect 4) is particularly interesting. Assume that the gate polysilicon is doped n-type and the gate oxide defect shorts the gate to the source of the n-channel transistor (VSS) by resistance R_{def}. Furthermore, let R_{on} and R_{off} be the conducting and non-conducting resistances of transistors (Fig. 5.21). In order to detect this defect, a logic high is written into the cell, i.e., node 'a' is high. Depending on the value of R_{def} with respect to R_{on} of p2 transistor, there are following possibilities:

- $R_{def} \ll R_{on}$ of p2 transistor: In this case node 'a' effectively is at SA0. Due to the positive feedback loop in the circuit, the p1 and n2 transistors are turned-on and n1 and p2 are turned-off. Hence, in the quiescent state no current flows and the defect is not detected by I_{DDQ}.

- $R_{def} \cong R_{on}$ of p2 transistor: In this case, the voltage levels of nodes 'a' and 'b' will depend upon the resistance ratio of the transistors. The intermediate voltage levels will give rise to high quiescent current making the defect I_{DDQ} detectable.

- $R_{def} \gg R_{on}$ of p2 transistor: Depending on the I_{DDQ} current threshold, the defect is detected or not detected. The current due to the defect can be approximately expressed as,

$$I \approx \frac{VDD}{R_{def}}$$
(EQ 5.3)

If the R_{def} is infinitely high (i.e., no defect) then the current is the defect-free normal leakage current of the transistor. However, if the current threshold is 10 μ A, then only a defect resistance that causes a quiescent current higher than that threshold is detected.

As explained in the previous subsection, many inter-cell bridging defects are not detected by the I_{DDQ} measurement. With a similar reasoning, it can be argued that a low resistance bridging defect causing a short between an internal node 'a' of a cell to node 'a' of an adjacent cell is also not detected by I_{DDQ}. Owing to such a defect a *write* operation affects both cells which have always the same logic value. Therefore no logic conflict is created.

5.7.2.2 DRAM Defects

Fig. 5.22(a) shows a part of a DRAM block diagram. The architecture and operation of a DRAM is quite similar to that of SRAM. The major difference between the two is in the data storage mechanisms. Unlike SRAMs, in DRAMs tiny capacitors hold the information in terms of charges. A typical value of such a small capacitor is 50 fF. The charge inherently leaks and therefore requires a periodic

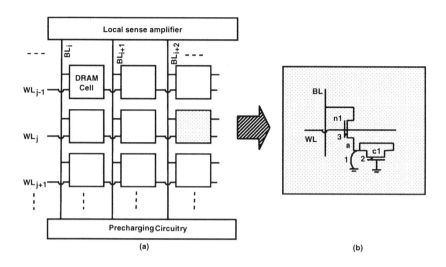

Fig. 5.22: (a) A part of DRAM block diagram, and (b) A DRAM cell and some of the I_{DDQ} undetectable defects.

refresh. Due to the dynamic storage of information, the application of an I_{DDQ} test method to a DRAM is less encouraging compared to an SRAM.

Fig. 5.22(b) shows a typical DRAM cell with some defects. It has an n-channel pass transistor and a capacitor to hold the tiny charge. With a similar reasoning as in the previous subsection, it can be concluded that all defects shown in Fig. 5.22(b) are not detected by I_{DDQ}. Moreover, a coupling fault due to a bridging defect affecting two or more cells is also not detected by the quiescent current measurement. Since the DRAM cell does not have any voltage strength or drive, it is not capable of creating and sustaining the quiescent current in case of such a defect. The stored charge in a DRAM cell is very small. Therefore, even a high impedance defect can seriously affect the storage capability. For example, a gate oxide defect (defect 2) causes charge on the capacitor to leak away. The approximate time constant, τ, of this leakage can be calculated as follows:

$$\tau = R_{def} \times C \qquad\qquad \text{(EQ 5.4)}$$

where, R_{def} and C are the gate oxide defect resistance and the DRAM cell storage capacitance, respectively. Assuming that the storage capacitance is 50 fF and the resistance value of the gate oxide defect is 10 kilo-ohms, simple calculation shows the resultant time constant, τ, as 10^{-11} seconds. Even a very high resistance gate oxide or bridging defect (say, 10^6 ohms) will give rise to a τ of 10^{-9} seconds. Moreover, the timing of a DRAM is asynchronously generated. In other words, like SRAM, DRAM defects in the present form do not give rise to elevated quiescent current and hence are not detected by the quiescent current test method.

5.7.3 I_{DDQ} Detection of RAM Defects

A careful examination suggests that I_{DDQ} testing is less efficient for RAMs due to the following reasons:

- In a RAM the internal timing is often generated by an asynchronous delay line.

- At any given moment there are several nodes that are in high impedance state.

- Since RAMs have a large number of bits, organized in unique addresses, I_{DDQ} measurements must be taken for addresses to have an effective defect coverage.

If the internal timing of a RAM is controlled such that a given wordline is activated and the corresponding bitline drives the given RAM cell for the complete duration of the I_{DDQ} measurement, then the intra-cell defects may be detected. For each RAM address at least two I_{DDQ} measurements are required with bitline driving logic high and logic low, respectively. This is illustrated for SRAMs as well as for DRAMs in Fig. 5.23. A large number of I_{DDQ} measurements, however, will make it impractical and expensive.

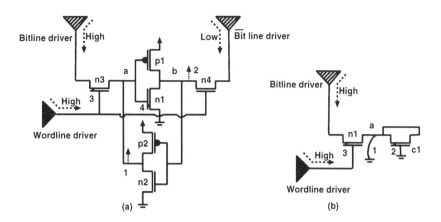

Fig. 5.23: The I_{DDQ} detection of (a) SRAM defects, and (b) DRAM defects by controlling the internal timing.

The number of I_{DDQ} samples required can be drastically reduced if in the test mode, complete RAM matrix is activated at the same time. All wordlines are activated and all bitlines are driven from bitline drivers. This arrangement ensures that all RAM cells are driven at the same time. Wordlines and bitlines are driven for the duration of the I_{DDQ} measurement. Bitlines are driven high and low and at both instances the I_{DDQ} measurements are taken. Therefore, only a few I_{DDQ} measurements are needed to detect most of the RAM matrix leakage defects.

However, a number of issues need to be addressed before such a concept can be implemented for an SRAM or a DRAM. For example, a large transient current will flow since all cells are activated in parallel. Bitline drivers must be modified to drive the increased load in the test mode. Power dissipation will be another issue and control circuit modifications will be needed to implement such an concept. These issues will be discussed in subsequent sections.

5.7.4 Test Complexity Reduction

March tests are popular because they allow economic and efficient RAM testing [4,39,51]. A typical march test algorithm performs a series of identical *read* and *write* operations for all addresses to cover a given fault model. Therefore, the test complexity of a march test is linear with respect to the number of RAM addresses. However, even linear operations when applied to multi-Mbit RAMs escalate the test cost substantially [17].

In this section, we demonstrate how efficient RAM test algorithms are obtained as a combination of the proposed I_{DDQ} testing method and voltage based march tests. The combined method is general enough to be applied to any CMOS RAM. As an example, we take a realistic fault model based on catastrophic process defects developed for SRAMs by Dekker et al. [4].

5.7.4.1 SRAM Fault Model and Test Algorithm

Following fault model proposed by Dekker et al. [4] is considered for this comparative study:

- A memory cell is stuck-at-0 or stuck-at-1.
- A memory cell is stuck-open.
- A memory cell suffers from a transition fault.
- A state of a memory cell is coupled to another cell.
- A memory cell has a multiple access fault.
- A memory cell suffers from a data retention fault in one (or both) of its states.

Dekker et al. proposed a basic algorithm of complexity 9N where N is the number of SRAM addresses. The algorithm covers the complete fault model except the data retention faults which require an additional 3N complexity [4]. The complexity of the basic algorithm rises to 13N if the read/write logic is sequential in nature. This algorithm is illustrated in Fig. 5.3. Word-oriented SRAMs require a careful consideration, since a single data background is not enough to ensure the detection of intra-word coupling faults. For example, in the case of a byte oriented SRAM four different data backgrounds may be required [4]. This means that the algorithm is to be run four times with different data backgrounds.

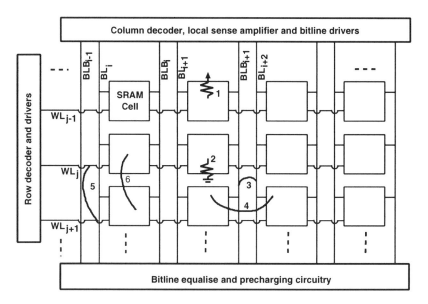

Fig. 5.24: Detection of SRAM defects by I_{DDQ}.

5.7.4.2 I_{DDQ} Defect Detection

A majority of the manufacturing process defects in RAMs give rise to catastrophic failures that can be easily detected [4,39]. Such defects in the modified SRAM are quickly detected by I_{DDQ} testing. The I_{DDQ} testing is a more
efficient method for the defect coverage. Therefore, in our combined test strategy, I_{DDQ} measurements are employed as the first pass/fail filter. Subsequently, march test is utilized for defects that are not detectable by I_{DDQ} testing and
for defects in address decoders. Defect detection with I_{DDQ} technique is further explained with the help of Fig. 5.24. This figure shows a simplified block
diagram of a SRAM with some commonly occurring faults in the cell matrix.
Let us assume that defects 1 and 2 cause two cells in the matrix to have stuck-
at-1 and 0 faults, respectively. Defect 3 causes a bridging fault between two
adjacent bit and $\overline{bit}$ lines. Defect 4 and defects 5 and 6 cause coupling faults
between adjacent horizontal and vertical cells, respectively.

For the first I_{DDQ} measurement, all wordlines are activated by wordline drivers and all bit and $\overline{bit}$ lines are driven to logic low and logic high, respectively,

from bitline drivers. Now the elevated quiescent current in this measurement will indicate the presence of a stuck-at-1 fault (defect 1) or a coupling fault (defect 3) in the cell matrix. For the second measurement, the logic values on bit and $\overline{bit}$ lines are complemented. Similarly, an elevated current in this measurement will indicate a stuck-at-0 fault (defect 2) or a coupling fault (defect 3) in the cell matrix. These two measurements together cover the gate oxide defects, leakage defects and the extra material defects causing data retention faults in SRAM cells. In order to detect defect 4, a pattern of 010101. is applied to successive bitlines with a complement pattern on $\overline{bit}$ lines. The presence of such a defect will cause elevated quiescent current in this measurement. Defect 5 causes a bridging defect between two successive wordlines. Therefore, it is detected by making all odd (or even) wordlines active and then taking the fourth I_{DDQ} measurement. It can be shown that third and fourth I_{DDQ} samples together will detect all state couplings proposed by Dekker et al. [4] except the state coupling involving two vertical cells (see defect 6, Fig. 5.24). Such a state coupling fault will be detected by the march test.

5.7.4.3 Undetected Defects

These I_{DDQ} measurements may not cover stuck-open faults in the matrix. Moreover, since address decoders are not exhaustively toggled, multiple access faults may not be sensitized for I_{DDQ} testing. Data retention faults may also not be exhaustively covered by I_{DDQ} measurements. For example, data retention faults caused by open defects will not be detected by such measurements. For these fault classes, the march test is best suited. The I_{DDQ} method is utilized to initialize the memory matrix before the first march element. After the initialization and in between marches, the SRAM is disabled for the data retention test. A bidirectional march is necessary for multiple access faults [4]. These two marches will also detect stuck-open faults and data retention faults. Fig. 5.25 shows the proposed test algorithm. The number of I_{DDQ} measurements is an absolute minimum for the coverage of the fault model with little or no diagnostic capability. Diagnostic capability is provided by the subsequent march test. Further diagnostics can be provided by additional features in the

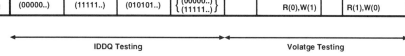

ADD	IDDQ1	IDDQ2	IDDQ3	IDDQ4			March1		March2
1				Wordlines selectively activated $\left\{ \begin{matrix} (010101..) \\ (101010..) \end{matrix} \right\}$			R(0),W(1)		R(1),W(0)
2	All wordlines activated (11111..)	All wordlines activated (11111..)	All wordlines activated (11111..)		Initialize 0 Disable RAM	Disable RAM	R(0),W(1)	Disable RAM	
.									
.							↘		↗
.									
.	All bitlines activated (00000..)	All bitlines activated (11111..)	All bitlines activated (010101..)	Bitlines selectively activated $\left\{ \begin{matrix} (00000..) \\ (11111..) \end{matrix} \right\}$					R(1),W(0)
N							R(0),W(1)		R(1),W(0)

← IDDQ Testing → ← Volatge Testing →

Fig. 5.25: Proposed SRAM test algorithm.

test mode. For example, the ability to selectively activate different rows and columns and then taking a I_{DDQ} sample will build additional diagnostic capability into the algorithm while increasing the algorithmic complexity. The number of samples depends on the size of the matrix and the degree of diagnostics required.

5.7.4.4 Algorithmic Complexity

The basic algorithm shown in Fig. 5.25 has the complexity of $4N + 4$ I_{DDQ} measurements. However, in the case of SRAM with sequential read/write logic, a three element march is necessary to test matrix stuck-open defects in the presence of output data latch [4]. Therefore, the first march is modified to include an additional R(1) after W(1) and the complexity of the proposed algorithm rises to $5N + 4$ I_{DDQ} measurements. Comparing algorithms in Fig. 5.3 and Fig. 5.25, we conclude that the complexity of the algorithm covering the given fault model is reduced substantially. Instead of 16N, the proposed algorithmic complexity is $5N + 4$ I_{DDQ} samples. The number of I_{DDQ} measurements is independent of the RAM complexity and two march elements are minimum for address decoder, stuck-open and data retention faults. As previously stated, the proposed methodology of combining I_{DDQ} testing and voltage based march tests can be applied to any CMOS RAM for similar benefits. For example, a similar analysis can be carried out with a DRAM algorithm based upon process defects [39] and the proposed method.

5.7.5 Implementation Aspects

The idea of activating the complete memory matrix and then taking I_{DDQ} measurements can be implemented in a number of ways. However, the implementation of such an idea for RAMs calls for a careful investigation. Various RAM circuit blocks are critically designed to meet performance, area and power dissipation constraints. There are several issues that should be addressed before the idea can be implemented. First of all, the DC current dissipating circuits either should be excluded or switched off in the I_{DDQ} test mode. For example, in DRAMs the substrate bias is often provided by a charge pump. The charge pump can create a high quiescent current state. For I_{DDQ} measurements such a pump should be shut-off.

A bitline driver is designed to drive only one cell in the normal operation. In the test mode, the same driver is inadequate to drive a large number of cells together within a given time. The problem is more complex in the case of SRAM where each cell is a tiny latch. A large transient current may flow if all cells are accessed together. This current flow should be suppressed to avoid electromigration and reliability problems. The flow of transient is effectively suppressed by turning-on wordlines sequentially in a binary fashion so that the bitline driver is not stressed. For example, at $t = 0$, the first wordline is enabled. Therefore, the first latch is overwritten. At $t = \delta$, next two wordlines are enabled. Now, the first latch helps the bitline driver to overwrite next two latches. Next, at $t = 2\delta$, four wordlines and then at $t = 3\delta$, eight wordline are enabled. In this manner, the previously overwritten latches help the bitline driver to overwrite the following latches on the bitline and very little transient current flows. Fig. 5.26 shows plots of transient current simulation for a SRAM having 64 wordlines. In the first case, when all the wordlines are enabled simultaneously, approximately 1.7mA transient current flows for 15 nanoseconds. In the second case, the wordlines are sequentially turned-on ($\delta = 2$ nanoseconds) and the flow of large transient current is suppressed. Testing a smaller number of wordlines in parallel (effectively increasing the number of I_{DDQ} measurements) is another solution. In DRAMs, the problem is manageable since the bitline drivers have to charge/discharge capacitors in-

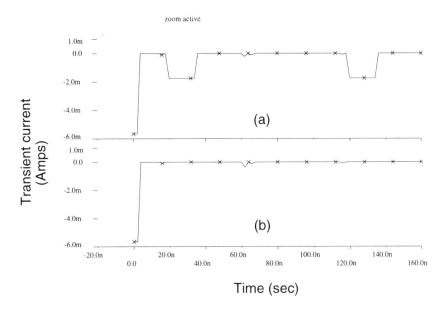

Fig 5.26: Transient current simulation for 64 wordline SRAM; (a) when all wordlines are enabled together, and (b) when they are turned on sequentially.

stead of overwriting the latches. However, a similar technique may be utilized for transient current suppression.

Fig. 5.27 shows a block diagram of the proposed SRAM. The memory control unit, precharge circuit and row and column decoders are modified to facilitate effective I_{DDQ} testing. The SRAM cell matrix and read/write logic remain unchanged. A test control signal, Test, is added to the SRAM controller for controlling the various conditions necessary for I_{DDQ} testing. When the Test is low, SRAM functions normally, however, a logic high on Test activates the test controller. Now the test controller controls different memory blocks in such a way that the necessary I_{DDQ} measurements are taken. The test control signals are shown with dotted lines.

A detailed implementation is beyond the scope of this book. In Fig. 5.28, we show how an existing wordline decoder can be modified to facilitate the I_{DDQ}

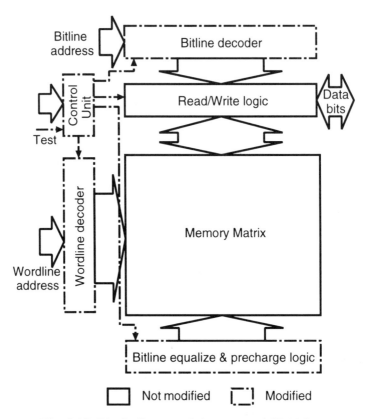

Fig. 5.27: Block diagram of the proposed SRAM.

testing. Other blocks can also be modified in a similar manner. Figure shows a 6 bit wordline decoder. In the unmodified decoder, a logic low on X_cntl forces all wordlines to be low (inactive state). The decoded address enables the corresponding wordline when X_cntl changes to logic high. In the test mode I_{DDQ} measurements require the row decoder to have the following additional functions: (a) to activate all wordlines, (b) to activate even numbered wordlines, and (c) to activate odd numbered wordlines at the same time. This is achieved by adding a two input AND gate before each wordline driver. The added gates are enclosed by a broken line in the figure. The even and odd numbered AND gates are controlled by two separate test control signals, namely Even_tst and Odd_tst. In order to turn on wordlines sequentially (for transient current suppression) logic delay elements (inverters) with different

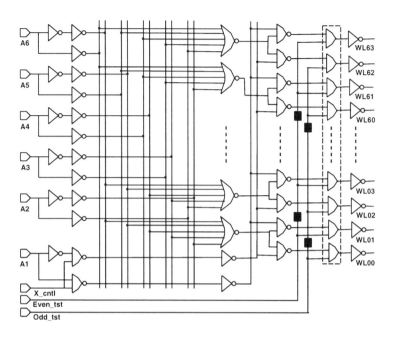

Fig. 5.28: Modified wordline address decoder to facilitate I_{DDQ} testing.

delays are added to the Even_tst and Odd_tst inputs of the AND gates. These delay elements are shown by black boxes.

Table 5.3 shows different operational modes of the row decoder. Logic low on both Even_tst and Odd_tst forces all wordlines to be activated. A low on Even_tst and a high on Odd_tst activates even numbered wordlines. However, the input row decoder address should also be kept even so as not to activate an additional odd numbered wordline. Similarly, a high on Even_tst and a low on Odd_tst activates odd numbered wordlines. In this case the input row decoder address should be kept as odd.

Such an implementation will not be without additional cost in area and in performance. However, extra logic required to implement the idea is small and fixed. The area increased is estimated between 3-4% of the total. It becomes even smaller as the RAM size increases. For future RAM generations, it will be of minor consequence. There is a possibility of increased access time be-

Table 5.3: Opretional modes of wordline address decoder.

Even_tst	Odd_tst	Mode
0	0	All wordlines activated
0	1	Even wordlines activated
1	0	Odd wordlines activated
1	1	Normal mode

cause there is an extra AND gate in the address decoder. However, the data path in RAMs is time critical and determines the access time. The data path is kept unchanged. The changes in control unit and bitline equalize circuitry will not affect the memory access time substantially. The worst case increase in access time is estimated as 2-3 nanoseconds. There should not be any increase in the normal mode power dissipation because extra logic is a small fraction of the total active devices.

5.8 Conclusion

RAMs enjoy a strategic position in the microelectronics industry and have been a cause of many trade battles. In terms of the volume, memories account for 20% of the semiconductor market [37]. As far as testing is concerned, RAMs suffer from quantitative issues of digital testing along with the qualitative issues of analog testing. Furthermore, RAM test cost and quality issues have become critical jeopardizing the development of future RAM generations.

Increasing miniaturization has forced RAMs to share the same substrate with CPU or DSP core. This merger has resulted in a dramatic change especially in the case of DRAMs which now must be fabricated by a process developed for standard logic. This leads to new challenges in the design and testing of embedded DRAMs. Defect oriented inductive fault analysis is carried out for an embedded DRAM module. Owing to the high circuit density, standard CMOS VLSI process implementation and dynamic nature of the operation, embedded DRAMs exhibit susceptibility to catastrophic as well as to non-catastrophic defects. The probability of occurrence of non-catastrophic defects is significantly high. Most of the non-catastrophic defects degrade the circuit performance and/or cause coupling faults in the memory.

The coupling faults in DRAMs need special attention. Previous definitions (transition based and state based) of coupling faults do not adequately represent their behavior in DRAMs. Such faults can be caused by catastrophic as well as non-catastrophic defects. Furthermore, they are dynamic in nature. A fault model is evolved taking into account the catastrophic and non-catastrophic defects and the dynamic nature of coupling faults. Based upon the fault model a test algorithm of complexity 8N is developed which completely covers the above mentioned fault model. The algorithm can be easily modified if the R/W logic is sequential in nature. The resultant algorithm has the complexity of 9N. For word oriented DRAMs, different backgrounds can be utilized to cover intra-word faults. However, if the bits constituting a word are not adjacent to each other in layout, then one data background is sufficient.

The effectiveness of these algorithms has been validated with tested DRAM devices. Most device failures (89%) appear to be total chip failures or memory stuck-at faults affecting a large number of bits. Such failures could easily be explained by catastrophic bridging or open defects at various locations in the layout. This result is consistent with the results reported by Dekker et al. [4]. Typically, a large number of bits failed due to such defects and, therefore, such failures were easily detected. The wafer yield was relatively low and hence many total chip failures occurred. A significant segment of failed devices (7%) showed bit stuck-at behavior that could also be explained with catastrophic defects in the cells. These defects caused relatively small number of bits to fail. However, a small number of device failures (4%) could not be explained with catastrophic defect model and could only be explained by foregoing coupling fault model based on non-catastrophic defects.

Much attention has been paid to contain the RAM test cost. Parallel RAM testing has become a standard practice for big (>1 Mbit) RAMs. MBT and LMT have gained acceptance for 4 Mbit and 16 Mbit DRAMs. Properties of linear feedback shift registers (LFSRs) have been exploited to generate test stimuli and compare the test response on chip to reduce the cost. Furthermore, attempts have been made to test many devices in parallel sharing the same tester resources, hence reducing the cost. Nevertheless, the implementation of any parallel RAM test scheme requires a careful consideration of several issues. High speed access as well as low power consumption must be maintained during parallel tests. Moreover, test time reduction should be significant, performance impact should be minimal, and area overhead of the scheme should be acceptable.

Arguably, I_{DDQ} testing is ideally suited for parallel testing because the fault information traverses in parallel over the power buses. Therefore, no special observability conditions are needed. This property of I_{DDQ} testing has been utilized to reduce the test costs of digital VLSI. However, a straightforward application of I_{DDQ} test technique to RAMs gives a very limited defect detection capability. Due to the very nature of the RAM architecture and operation, the I_{DDQ} testing is not able to detect many of the otherwise I_{DDQ} detectable manufacturing process defects. With minor design modifications, the I_{DDQ} testing is made a potent method for the detection of bridging and gate oxide defects in CMOS RAMs. Since, I_{DDQ} testing can be utilized to test a large number of cells in parallel, it results in efficient testing of manufacturing process defects in RAM cell matrix. At the same time, it can not robustly detect faults caused by open defects. Address decoder faults and the sequential nature of output data can be best tested by voltage based march tests. We evolved RAM test algorithms considering I_{DDQ} testing as well as voltage based march test. The effectiveness of the proposed methodology is demonstrated with a realistic SRAM fault model. The combined test methodology reduces the algorithmic test complexity from 16N to 5N + 4 I_{DDQ} measurements. The methodology is general enough to be applied to any CMOS RAM.

Like any other parallel test scheme for RAMs, the implementation of the I_{DDQ} based parallel testing of RAMs also requires a substantial effort. Various RAM circuit blocks are critically designed to meet performance, area and power dissipation requirements. DRAMs often have substrate charge pumps and sense amplifiers generating high quiescent currents. These have to be in-activated for I_{DDQ} testing. Furthermore, modifications in address decoders and control unit are required. These modifications should be such that their impact on circuit performance and reliability is minimal. The modifications cost additional area (3-4%) and are likely to affect the performance (worst case access time degradation is 2-3 nanoseconds). As the complexity of RAM under test increases the proposed test scheme becomes more attractive.

Despite the implementation effort, and area and performance penalties, the proposed method is very attractive for overall RAM test cost reduction while ensuring better product quality. Neither I_{DDQ} nor voltage testing alone can en-sure the quality requirements of the industry. The area cost is almost constant and is a very small fraction of the total area. The performance degradation will strongly depend on the implementation. Since this concept can be incorporat-ed in a number of ways, those with minimal performance implications can be chosen to minimize the performance degradation for high performance appli-cations.

References

1. M.S. Abadir and H.K. Reghbati, "Functional Testing of Semiconductor Random Access Memories," *ACM Computing Surveys*, vol. 25, no.3, pp. 175-198, September 1983.

2. M.A. Breuer and A.D. Friedman, *Diagnosis and Reliable Design of Digital Systems,* Rockville, MD: Computer Science Press, 1976.

3. E.M.J.G. Bruls, "Reliability Aspects of Defects Analysis," *Proceedings of European Test Conference,* 1993, pp. 17-26.

4. R. Dekker, F. Beenker and L. Thijssen, "Fault Modeling and Test Algorithm Development for Static Random Access Memories," *Proceedings of International Test Conference*, 1988, pp. 343-352.

5. B.N. Dostie, A. Silburt and V.K. Agarwal, "Serial Interfacing for Embedded-Memory Testing," *IEEE Design & Test of Computers*, vol. 7, pp. 52-63, April 1990.

6. F.J. Ferguson, and J.P. Shen, "Extraction and simulation of realistic CMOS faults using inductive fault analysis," *Proceedings of International Test Conference*, 1988, pp. 475-484.

7. A.V. Ferris-Prabhu, "Computation of the critical area in semiconductor yield theory," *Proceedings of the European Conference on Electronic Design Automation*, 1984, pp. 171-173

8. M. Franklin, K.K. Saluja and K. Kinoshita, "Design of a BIST RAM with Row/Column Pattern Sensitive Fault Detection Capability," *Proceedings of International Test Conference*, 1989, pp. 327-336.

9. M. Franklin, K.K. Saluja and K. Kinoshita, "Row/Column Pattern Sensitive Fault Detection in RAMs via Built-in Self-Test," *Proceedings of Fault Tolerant Computing Symposium*, June 1989, pp. 36-43.

10. M. Franklin, K.K. Saluja, and K. Kinoshita, "A Built-In Self-Test Algorithm for Row/Column Pattern Sensitive Faults in RAMs," *IEEE Journal of Solid State Circuits*, vol. 25, no. 2, pp. 514-523, April 1990.

11. A.J. van de Goor, *Testing Semiconductor Memories, Theory and Practice,* John Wiley and Sons, 1991.

12. T. Guckert, P. Schani, M. Philips, M. Seeley and H. Herr, "Design and Process Issues for Elimination of Device Failures Due to 'Drooping' Vias," *Proceedings of International Symposium for Testing and Failure Analysis (ISTFA)*, 1991, pp. 443-451.

13. J.P. Hayes, "Detection of Pattern-Sensitive Faults in Random Access Memories," *IEEE Transactions on Computers,* vol. C-24, no.2, pp. 150-157, February 1975.

14. J.P. Hayes, "Testing Memories for Single-Cell Pattern-Sensitive Faults," *IEEE Transactions on Computers*, vol. C-29, no.3, pp. 249-254, March 1980.

15. L. K. Horning, J.M. Soden, R.R. Fritzemeier and C.F. Hawkins, "Measurements of Quiescent Power Supply Current for CMOS ICs in Production Testing," *Proceedings of International Test Conference,* 1987, pp. 300-309.

16. J. Inoue, T. Matsumura, M. Tanno and J. Yamada, "Parallel Testing Technology for VLSI Memories," *Proceedings of International Test Conference*, 1987, pp. 1066-1071.

17. M. Inoue, T. Yamada and A. Fujiwara, "A New Testing Acceleration Chip for Low-Cost Memory Test," *IEEE Design & Test of computers*, vol. 10, pp. 15-19, March 1993.

18. J.Knaizuk and C.R.P. Hartman, "An Optimal Algorithm for Testing Stuck-At Faults in Random Access Memories," *IEEE Transactions on Computers*, vol. C-26, no. 11, pp. 1141-1144, November 1977.

19. S. Koeppe, "Optimal layout to avoid CMOS stuck-open faults," *Proceedings of 24th Design Automation Conference,* 1987, pp. 829-835.

20. M. Kumanoya, et al., "A 90ns 1Mb DRAM with Multi-Bit Test Mode," *International Solid State Circuits Conference; Digest of Technical Papers*, 1985, pp. 240-241.

21. K.J. Lee and M.A. Breuer, "Design and Test Rules for CMOS Circuits to Facilitate I_{DDQ} Testing of Bridging Faults," *IEEE Transactions on Computer-Aided Design*, vol. 11, no. 5, pp. 659-669, May 1992.

22. M.E. Levitt and J.A. Abraham, "Physical Design of Testable VLSI: Techniques and Experiments," *IEEE Journal of Solid State Circuits*, vol. 25, no. 2, pp. 474-481, April 1990.

23. W. Maly, "Realistic Fault Modeling for VLSI Testing," *Proceedings of 24th ACM/IEEE Design Automation Conference*, 1987, pp.173-180.

24. W. Maly and M. Patyra, "Design of ICs Applying Built-in Current Testing," *Journal of Electronic Testing: Theory and Applications*, vol. 3, pp. 397-406, November 1992.

25. Y. Matsuda, et al., "A New Parallel Array Architecture For Parallel Testing in VLSI Memories," *Proceedings of International Test Conference*, 1989, pp. 322-326.

26. P. Mazumder, "Parallel Testing of Parametric Faults in Three Dimensional Random Access Memory," *IEEE Journal of Solid State Circuits*, vol. SC-23, pp. 933-941, 1988.

27. P. Mazumder and K. Chakraborty, *Testing and Testable Design of High-Density Random-Access Memories,* Boston: Kluwer Academic Publishers, 1996.

28. H. McAdams, et al., "A 1-Mbit CMOS Dynamic RAM with Design For Test Functions," *IEEE Journal of Solid State Circuits*, vol. SC-21, pp. 635-641, October 1986.

29. R. Meershoek, B. Verhelst, R. McInerney and L. Thijssen, "Functional and Iddq Testing on a Static RAM," *Proceedings of International Test Conference*, 1990, pp. 929-937.

30. A. Meixner and W. Maly, "Fault Modeling for the Testing of Mixed Integrated Circuits," *Proceedings of International Test Conference*, 1991, pp. 564-572.

31. S. Naik, F. Agricola and W. Maly, "Failure analysis of High Density CMOS SRAMs Using Realistic Defect Modeling and I_{DDQ} Testing," *IEEE Design & Test of Computers,* vol. 10, pp. 13-23, June 1993.

32. R. Nair, "Comments on an Optimal Algorithm for Testing Stuck-at Faults in Random Access Memories," *IEEE Transactions on Computers*, vol. C-28, no. 3, pp. 258-261, March 1979.

33. R. Nair, S.M. Thatte and J.A. Abraham, "Efficient Algorithms for Testing Semiconductor Random Access Memories," *IEEE Transactions on Computers*, vol. C-27, no. 6, pp. 572-576, June 1978.

34. H.D. Oberle and P. Muhmenthaler, "Test Pattern-Development and Evaluation for DRAMs with Fault Simulator RAMSIM," *Proceedings of International Test Conference*, 1991, pp. 548-555.

35. C.A. Papachristou and N.B. Sahgal, "An Improved Method for Detecting Functional Faults in Semiconductor Random Access Memories," *IEEE Transactions on Computers*, vol. C-34, no.2, pp. 110-116, February 1985.

36. R. Perry, "I_{DDQ} testing in CMOS digital ASICs," *Journal of Electronic Testing: Theory and Applications*, vol. 3, pp. 317-325, November 1992.

37. B. Prince, *Semiconductor Memories,* Chichester, UK: John Wiley and Sons, 1991.

38. M. A. Rich and D. E. Gentry, "The Economics of Parallel Testing," *Proceedings of International Test Conference*, 1983, pp. 728-737.

39. M. Sachdev and M. Verstraelen, "Development of a Fault Model and Test Algorithms for Embedded DRAMs," *Proceedings of the International Test Conference*, 1993, pp. 815-824.

40. M. Sachdev, "Transforming Sequential Logic in Digital CMOS ICs for Voltage and I_{DDQ} Testing," *Proceedings of European Design and Test Conference,* 1994, pp. 361-365.

41. M. Sachdev, "Reducing the CMOS RAM Test Complexity with I_{DDQ} and Voltage Testing," *Journal of Electronic Testing: Theory and Applications (JETTA),* vol. 6, no. 2, pp. 191-202, April 1995.

42. J. Savir, W.H. McAnney and S.R. Vecchio, "Testing for Coupled Cells in Random Access Memories," *Proceedings of International Test Conference*, 1989, pp. 439-451.

43. Semiconductor Industry Association (SIA), "*The National Technology Roadmap for Semiconductors,*" pp. 94-99, 1994.

44. A. H. Shah, et al., "A 4-Mbit DRAM with Trench Transistor Cell," *IEEE Journal of Solid State Circuits,* vol. SC-21, pp. 618-627, October 1986.

45. J.P. Shen, W. Maly and F.J. Ferguson, "Inductive Fault Analysis of MOS Integrated Circuits," *IEEE Design & Test of Computers*, vol. 2, no. 6, pp. 13-26, 1985.

46. J.M. Soden, C.F. Hawkins, R.K. Gulati and W. Mao, "I_{DDQ} Testing: A Review," *Journal of Electronic Testing: Theory and Applications*, vol. 3, pp. 291-303, November 1992.

47. T. Sridhar, "A New Parallel Test Approach for Large Memories," *Proceedings of International Test Conference*, 1985, pp. 462-470.

48. F.A. Steenhof, C.G. van der Sanden and B.C. Pham, "Design Principles of a DRAM Cell Matrix for Embedded Applications," *Nat.Lab. internal technical note,* TN 250/90.

49. S.T. Su and R.Z. Makki, "Testing of Static Random Access Memories by Monitoring Dynamic Power Supply Current," *Journal of Electronic Testing: Theory and Applications*, vol. 3, pp. 265-278, August 1992.

50. D.S. Suk and S.M. Reddy, "Test Procedure for a Class of Pattern-Sensitive Faults in Random Access Memories," *IEEE Transactions on Computers*, vol. C-29, no.3, pp. 419-429, June 1980.

51. D.S. Suk and S.M. Reddy, "A March Test for Functional Faults in Semiconductor Random Access Memories," *IEEE Transactions on Computers*, vol. C-30, no.12, pp. 982-985, December 1981.

52. M. Syrzycki, "Modeling of Spot Defects in MOS Transistors," *Proceedings of International Test Conference*, 1987, pp. 148-157.

53. E. Takeda, et al., "VLSI Reliability Challenges: From Device Physics to Wafer Scale Systems," *Proceedings of IEEE*, vol. 81, no. 5, 1993, pp. 653-674.

54. S.M. Thatte and J.A. Abraham, "Testing of Semiconductor Random Access Memories," *Proceedings of Fault Tolerant Computing Symposium*, 1977, pp. 81-87.

55. H. Walker and S.W. Director, "VLASIC: A Catastrophic Fault Yield Simulator for Integrated Circuits," *IEEE Transactions on Computer Aided Design of Integrated Circuits and Systems*, vol. 5, no. 4, pp. 541-556, 1986.

56. H. Yokoyama, H. Tamamoto and Y. Narita, "A Current Testing for CMOS Static RAMs," *Proceedings of IEEE International Workshop on Memory Technology, Design and Testing*, August 1993, pp. 137-142.

57. Y. You and J.P. Hayes, "A Self Testing Dynamic RAM Chip," *IEEE Journal of Solid State Circuits*, vol. SC-20, no.1, pp. 428-435, February 1985.

CHAPTER 6

Testing Defects in Programmable Logic Circuits

Recent resurgence of programmable logic devices (PLDs) ushers a new era in digital system design. Modern PLD architectures offer flexibility of logic implementation. However, the same flexibility results in enormous test complexity. In this chapter, a brief review of testable PLD architecture is presented and methodologies for testing defects are outlined.

6.1 Introduction

Programmable logic is widely used in VLSI circuits and systems since it provides a simple method of implementing complex functions. These devices range from simple programmable logic arrays (PLAs) to high complexity field programmable gate arrays (FPGAs) and complex programmable logic devices (CPLDs). In its simplest form a PLA is a highly uniform structure capable of implementing any Boolean function expressed as a sum of products. As illustrated in Fig. 6.1 a PLA structure contains an AND-plane and an OR-plane. The AND-plane uses wired logic. Each AND-plane output is pulled "high" by a pull-up transistor (or resistor) and can be pulled "low" by any input connected to it. Similarly, the OR-plane is also built from wired logic.

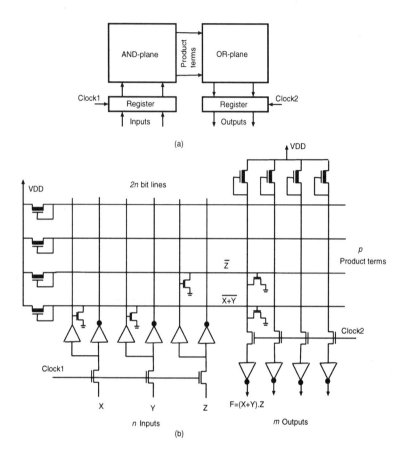

Fig: 6.1: Block diagram and architecture of an NMOS PLA.

According to De Morgan's rule a sum of products can be realized in several ways. For example, a Boolean function given in the sum-of-product form can also be implemented by either NOR-NOR planes, or NAND-NOR planes, or NAND-NAND planes.

The block diagram of a PLA is illustrated in Fig. 6.1(a). It has n inputs, p product lines, and m outputs. Each outputs is a Boolean function realized by the PLA. Fig. 6.1(b) illustrates the implementation of a PLA in NMOS technology. The AND-plane has $2n$ columns called bit lines, and p rows corresponding to p product terms. Each bit line carries a logic input or its complement as determined by a single bit decoder. The product lines are pulled up by depletion mode pullup transistors. The desired product term function is realized by placing switches on the intersection of appropriate bit lines with the product line. Each product term is formed by a wired NOR logic operation of selected bit lines. The OR-plane is also realized through wired NOR logic operation. The product lines from the AND-plane traverse through and form the inputs to the OR-plane. The m outputs lines form the columns. The sum of products on a particular output line is realized by putting switches on the intersection of appropriate product terms with the output line. This type of PLA is also known as a NOR-NOR PLA.

6.2 Evolution of Programmable Circuits

In this section, we briefly review the evolution of programmable logic archi-
tectures. For a detailed treatment an interested reader is referred to available textbooks on the subject [4,32].

Programmable read only memories (PROMs) were the first devices in which a user could implement a Boolean function. Address lines are used as logic circuit inputs and data lines as outputs. However, PROMs were very inefficient for implementing logic functions. Still in a true sense, they were the first devices to incorporate user programmability.

However, PLAs were the first programmable devices that offered flexibility to designers. As mentioned before, it had a one-time programmable AND-plane as well as a one-time programmable OR-plane. Two levels of programmability made it possible to implement any Boolean function in the sum of products form. PLAs became popular with chip designers as well as with system de-
signers. At the chip level, PLAs are often programmed with a contact mask. Alternatively, they are specifically designed to realize a particular function which is not programmable. PLAs at the chip level are commonly used to implement instruction decoders of microprocessors, and combinational circuitry of finite-state machines. Their popularity comes from the fact they are simple

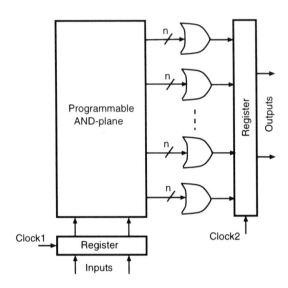

Fig. 6.2: A genralized PAL block diagram.

to design owing to their regular architecture, and the availability of synthesis
programs that automate the synthesis process.

At the board or system level, user programmability offers flexibility of quick
prototyping, last minute reconfiguration, etc. Therefore, these devices are sold
as dedicated, user programmable chips. However, two levels of configurable
architecture gave rise to poor speed performance and expensive manufacturing costs. Therefore, the PLA architecture gave way to programmable array
logic (PAL) architecture. Fig. 6.2 illustrates a generalized PAL architecture. It
has only a single level of programmability, wired AND-plane, and has a fixed
OR gates. Single level of programmability resulted in some constraints on arbitrary logic implementations. However, in spite of the fixed OR-plane, usually, high percentage of gates can be utilized. PAL devices come in a number of
configurations, with different numbers and different sizes of OR gates. In order to implement sequential circuits, usually flip-flops (registers) are added at
the inputs and outputs of a PAL.

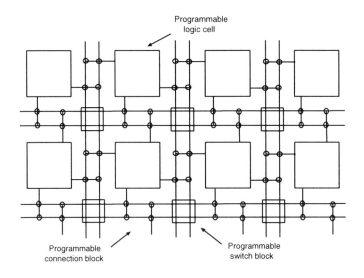

Fig. 6.3: A typical field programmable gate array architecture.

Introduction of PAL devices brought a paradigm shift in the way digital system designs are implemented. A number of recent architectures have evolved based upon the PAL concept. Brown and Rose [5] collectively named them as simple programmable logic devices (SPLDs). These devices are commercially available from several vendors. However, advances in technology created a need for higher density PLDs. Pure SPLD architecture has the disadvantage that the programmable plane structure grows very quickly as the number of inputs is increased. One way to alleviate this problem is to have a programmable interconnect matrix that connects individual SPLDs on the same chip. Such devices are known as complex programmable logic devices (CPLDs). The programmability is achieved through SRAM cells, or EEPROMs, or one time programmable (OTP) switches, etc.

Field programmable gate arrays (FPGAs) are another popular way to quickly implement random logic. From an architectural point of view FPGAs are different from CPLDs. The architecture of FPGAs evolves from mask programmable gate arrays (MPGAs). An FPGA contains an array of programmable logic elements interspersed by a programmable interconnect network. The interconnect network consists of programmable switches that are organized in connection blocks and switching blocks. These switches are MOS transistors programmed (conducting or non-conducting behavior) by SRAM cells, anti-fuses, or (E)EPROMs.

Fig. 6.3 illustrates important building blocks of an FPGA. Although FPGAs and SPLDs both are programmed in a similar manner, there are important differences between the two. FPGAs can achieve much higher level of integration than SPLDs. However, recent developments on CPLDs aim at a similar level of integration. The routing strategies in a FPGA are similar to that of a conventional VLSI where each connection typically passes through several switches. In a SPLD routing architecture is very simple but highly inefficient. Every output is directly connectable to every input through one switch. Furthermore, in SPLDs, logic is often implemented with only two level AND-OR logic. In other words, a logic function is transformed into an equivalent high fan-in two level form which is implemented in the SPLD. On the other hand, in an FPGA logic is implemented using multiple levels of small fan-in gates.

6.3 PLA Test Complexity

Modern PLDs have a programmable architecture that includes logic, SRAM, EEPROM, and a large number of programmable interconnects on the same substrate. Such complex devices are difficult to model using conventional fault modeling techniques. Another distinctive feature of modern PLDs is the re-programmability that adds to test complexity. Early PLA and PAL devices were one time programmable (OTP) devices, either personalized at fab or by the end-user. Therefore, a test for those devices could be based on the implemented function. However, re-programmability necessitates a test for total functionality including the programmable devices. Therefore, there have been efforts to determine the test that is independent of the function implemented (universal test) in the PLA [1].

In order to appreciate the test complexity of modern programmable logic devices, it is important to review the basic PLA test issues. We focus on PLAs because they are still popular in embedded applications compared to the PAL architecture. Moreover, the test complexity of PAL devices with only one programmable plane is lower than that of PLAs. Furthermore, in this book, we are primarily concerned with testing of defects in embedded building blocks of an IC. Finally, the basics of PLA testing have relevance in the testing of present day CPLD and FPGA architectures as well. An overview of PLA test issues and easily testable PLA design techniques has been given by Agarwal [1].

6.3.1 PLA fault models

PLA fault models include line SA faults, bridging faults, open faults, and missing and extra crosspoint faults. In user-programmable PLAs, each crosspoint can be programmed as off or on. Therefore, in these PLAs programmability of crosspoint switches needs to be tested instead of an extra or missing crosspoint fault. Typical PLA faults consist of following individual fault types:

- **Line SA faults:** Single or multiple line SA faults are considered under this category. The faults are considered on input lines, product lines, output lines, input/output registers, and pullup logic.

- **Crosspoint faults:** A crosspoint may exist at an undesired location, or a crosspoint may be missing from a desired location. These faults are known as extra and missing crosspoint faults, respectively. These faults are also known as growth and shrinkage faults. In modern PLDs, a (re)-programmable switch is used to define the presence or absence of a crosspoint at a particular location. Testing for programmability should cover the crosspoint faults.

- **Open faults:** are considered on input lines, product lines, and output lines. Most opens in PLA occur on lines or in single transistors and lead to SAF behavior on lines or in pullup (pulldown) transistors. Therefore, these need not to be considered explicitly. The early work on the PLA fault model development did not take bridging faults into account since for early NMOS (CMOS) process bridging defects were less likely because packing densities were not as high. Arguably, these faults are the most frequently occurring faults in the current CMOS technology. For a PLA, the bridging faults may be defined as:

- **Bridging faults:** Bridging faults among input lines, product lines, and output lines are considered. Furthermore bridging faults between input and product lines, product lines and output lines are also considered.

Crosspoint faults (cp) need special attention. In general, a PLA with n input lines, p product lines, and m output lines, can have as many as $(2n + m)p$ single cp faults and $2^{(2n + m)p} - 1$ single and multiple cp faults. However, a test designed to detect all single cp faults in a PLA covers most multiple cp faults. Moreover, such a test is also shown to cover most SA, open, and short faults [2,22,29]. Nevertheless, finding a complete test for all single cp faults in a given PLA is a computationally difficult task in spite of various heuristic-based algorithms [2,21,22,23,29]. The problem is further compounded by the fact that almost all PLAs contain inherently untestable cp faults. A cp fault is said to be untestable if the logical functions realized by the faulty and the fault-free PLAs are identical.

6.3.2 Testability Criterion for PLAs with DfT

The complex problem of test generation provided motivation for making PLA architectures testable. Since PLAs have a very regular architecture, there are several ways of making them easily testable. A substantial research effort is spent on devising testable architectures and finding out their trade-offs in terms of overheads [1]. Although design and test complexities of modern PLAs is more complex than the simple early PLAs, fundamental test issues have not changed. Therefore, it is important to review testable PLA architectures. In order to quantify the benefits of testability schemes for PLAs. Agarwal [1] described five important criterion:

- **Fault Coverage:** Generally, it is the ratio of single cp faults covered by the DfT scheme to the total number of possible cp faults $\{(2n + m)p\}$. However, it could also be fault coverage in terms of SA faults.

- **Number of Test Patterns:** The total number of test patterns needed to test all single cp faults.

- **Function Independence:** If the test patterns for a PLA are developed without prior knowledge of the function realized by the PLA, then the corresponding test set is said to be function independent or universal. Such a test is desirable for modern PLAs and PLDs that may be programmed many times by the user for diverse functions.

- **Extra Hardware:** Testability schemes require extra logic, interconnects in order to achieve full cp fault testability.

- **Test Time Delay:** This is calculated by multiplying the number of test patterns used by the delay per test pattern. The latter quantity is obtained by assuming that the input decoder, the AND-plane, and the OR-plane, each introduces delay. The maximal delay experienced by a test pattern from input to the output in the test mode is referred to as the delay per test.

6.4 Testability Schemes for PLAs

A brute-force test generation method for PLAs is more complex than that for random logic due to a variety of reasons. PLAs contain logic redundancies that may produce untestable faults. Secondly, a PLA contains a significant amount of reconvergent fanout that makes the task of test generation difficult. Finally, an embedded PLA has interdependencies among its inputs that may exclude the input pattern needed to stimulate and propagate the fault effect through the PLA [19]. On the other hand, the structural organization of PLA makes test generation easier if some kind of testability scheme is incorporated. The key to the design of easily testable PLA is the independent control of input and product lines. By changing the input patterns, the input lines are changed in pairs and not individually. Similarly, input patterns may change more than one product lines at a time hence may result in fault masking or undetected cp faults.

The first design of PLAs with product and bit line selection capability was independently proposed by Hong and Ostapko [14] and Fujiwara et al. [8,9].

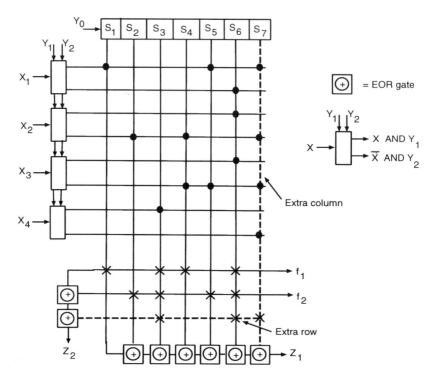

Fig. 6.4: Augmented PLA proposed by Fujiwara et al. [12,13].

Both schemes proposed the use of extra hardware to select a given bit line of the matrix, a shift register to select any given product line, and two EXCLUSIVE-OR cascades or trees to check the parity of the bit and product lines. Such easily testable PLAs were known as augmented PLAs. An important feature of this design was that the test patterns and response were independent of the function implemented by the PLA for single as well as multiple SA and cp faults. In other words, the test patterns as well as response were universal for a given size (n, p, m) of the PLA. Furthermore, since the augmented PLAs have the universal test set, the test generation of the PLA is no longer necessary. Yajima and Aramaki [34] utilized this scheme to make a built-in self testable (BIST) PLA. Fig. 6.4 illustrates the testability technique proposed by Fujiwara et al. [8,9]. The scheme proposed by Hong and Ostapko [14] has the disadvantage that it needs more hardware, however, needs less test time compared to the one proposed by Fujiwara et al.

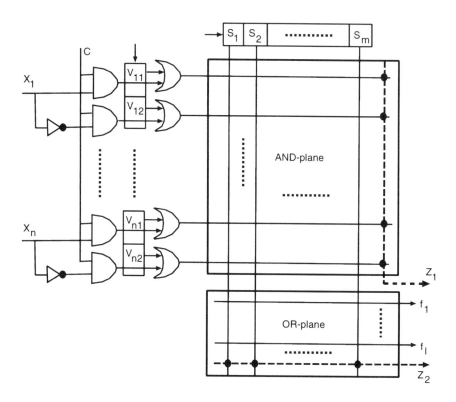

Fig. 6.5: Augmented PLA proposed by Saluja et al. [15,16].

Saluja et al. [27,28] described an easily testable PLA design with high fault coverage and low area overhead. The block diagram of the schemes is shown in Fig. 6.5. They replaced two cascades of EXCLUSIVE-OR gates with an additional shift register between the decoder and the AND-plane resulting in reduced area overhead. The PLA had a universal test set, capable of detecting multiple stuck-at and cross point faults. However, the major disadvantage of the scheme was extra gate delays in the PLA data path making it slower in normal operation. Although the test patterns were independent of the implemented function, the output responses were not function independent which made application of BIST difficult. Finally, Saluja et al. [27,28] did not consider bridging faults which are an important faults for PLAs and CMOS circuits.

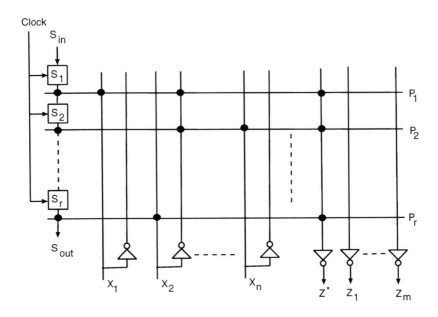

Fig. 6.6: The easily testable PLA architecture proposed by Khakbaz.

In a totally different approach Khakbaz [16,17] suggested inclusion of a shift register to control individual product lines and an extra row in the OR-plane. The scheme is illustrated in Fig. 6.6. This additional row is simply the OR of all product lines, which in combination with the shift register can be used to observe changes on any product line without masking. There is no additional control needed on the input decoder. Instead, all cp faults on a given product line are tested by first activating that product line by the shift register, and then supplying any input pattern that will produce a 1 on the product line. The testing starts by applying a sequence of input patterns beginning with the pattern selected above, each pattern being different from the previous one in one input variable considered thus far. Each change either affects the product term, in which case it is observed on the additional row, or it is a don't care, in which case no change is expected on the additional row. By exhaustively sensitizing each crosspoint, it is possible to detect all single and multiple cp faults in the AND-plane. The OR-plane faults are detected by activating one product line at a time, and by reading all the output lines. The shift register can be tested in the scan mode.

6.5 BIST for PLAs

Built-in self test (BIST) is very attractive to alleviate the test complexity of PLAs. BIST PLAs are so attractive that they have been used in some of the commercial microprocessors [12,18]. The concept of BIST PLA was independently proposed by Yajima and Aramaki [34] and Daehn and Mucha [7] and later improved by others [10,13,15]. Area overhead and fault coverage are the most important criterion for BIST PLA. These two criteria are often contradictory to each other. Furthermore, while designing a practical BIST PLA, care should be taken so that test patterns as well as test response are independent of the function implemented. Many of the PLA testability schemes mentioned in the previous section can be implemented with a BIST approach. However, a testability scheme that does not produce a function independent output response for the applied test patterns may not be suitable for a cost effective BIST application.

Fig. 6.7 depicts a block diagram of a BIST PLA proposed by Yajima and Aramaki [34]. The additional hardware consists of (i) four additional product terms, (ii) two additional output lines, (iii) parity circuits on AND and OR planes, (iv) a shift register as product term selector, (v) modified input decoder with a shift register, (vi) feedback value generator (FVG), and (vii) a test control signal. All the additional hardware is shown with broken lines in the figure.

The test control signal activates the test. The initial test pattern is scanned into the input shift register and product term selector. The application of the clock results in parity values from AND and OR planes. These parity signals with last two of the four additional product terms are fed into the FVG which outputs Y to generate the next test vector. In this manner, all test patterns are applied. The generated test patterns are independent of the function implemented in the PLA. If there is a fault that creates a faulty response, it

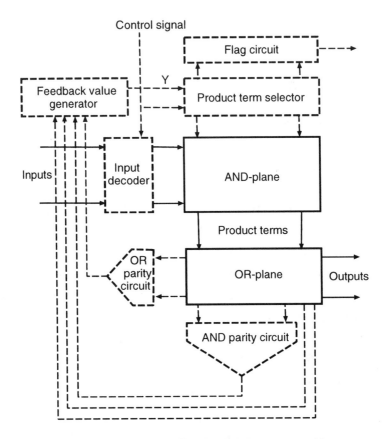

Fig. 6.7 BIST PLA proposed by Yamija and Aramaki.

will result in a faulty value in the product term selector which can be scanned out for a pass/fail decision. A total of $n+2p+8$ test vectors are needed for the test. This scheme covers all single SA faults, all cp faults, and all SA faults affecting the BIST circuitry. However, the coverage of multiple faults is not guaranteed.

The concept of built in logic block observer (BILBO) for BIST PLA was proposed by Daehn and Mucha [7]. BILBO is a multi-purpose building block that can be used as (i) Pseudo random test pattern generator by configuring as a linear feedback shift register, (ii) Signature compression circuit by configuring as multi input linear feedback shift register, (iii) A normal shift register, and

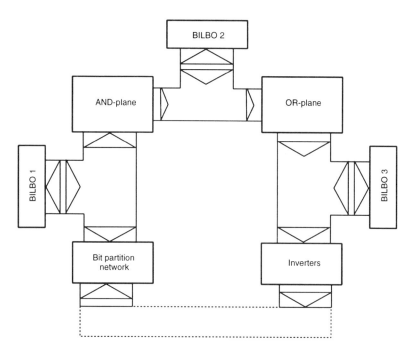

Fig. 6.8: BILBO based BIST scheme for PLAs.

(iv) Individual latches. Daehn and Mucha demonstrated the use of versatile BILBOs to implement a BIST PLA. Their scheme is illustrated in Fig. 6.8.

Three BILBOs are used to generate complete tests for detecting all faults in a NOR plane. The test has three distinctive consecutive steps.

- **Test for first NOR plane:** The test patterns are generated by BILBO 1 and test response is evaluated by BILBO 2. Subsequently, the content of BILBO 2 is shifted out for a pass/fail decision.

- **Test for second NOR plane:** Similar to the previous one, the test patterns are generated by BILBO 2 and the test responses are evaluated by BILBO 3. Subsequently, the content of BILBO 2 is shifted out for a pass/fail decision.

- **Test for output inverters and Bit partitioning network:** These parts are tested by feeding back the primary output to the primary input. For this

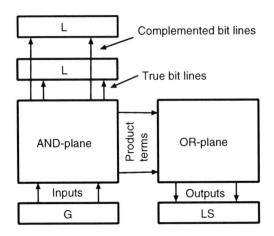

Fig. 6.9: BIST PLA proposed by Hassan and McCluskey

purpose, BILBO 3 is used as an input pattern generator and BILBO 1 is used for pattern evaluation.

This BIST scheme offered good fault coverage and needed only $n+p+m$ test vectors. However, for BIST application, the test vector length is of minor consequence compared to the area overhead which is substantially large in this case. Although the test vectors are function independent, the output response is not function independent. Furthermore, both schemes are suitable for small PLAs because for large PLAs the overheads become too large, and fault coverage becomes low [10].

An exhaustive test scheme using signature analyzers was proposed by Hassan and McCluskey [13]. The scheme is illustrated in Fig. 6.9. All possible pattern combinations except all zeros are applied by means of an LFSR, G, to the PLA and the output response of the PLA is evaluated by three different MISRs. These MISRs were placed on the true and complement bit lines and on the output of the PLA. A fault is said to be detected if the data in an MISR after the test is different from the fault free expected data. These MISRs augment the observability of the PLA and alleviate the problem of reconvergent

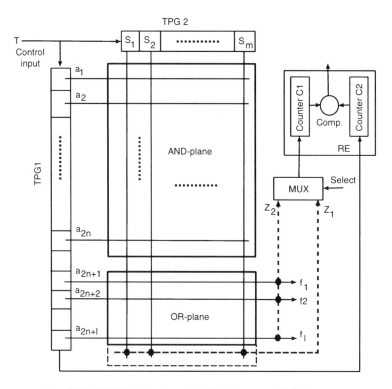

Fig. 6.10: BIST PLA scheme by Upadhaya and Saluja.

fanouts. Therefore, such a scheme can detect redundant faults as well as some multiple faults. However, exhaustive nature of test patterns make it impractical for large PLAs.

Hua et al. [15] suggested a BIST implementation of the PLA testability scheme proposed by Fujiwara et al. [8,9]. They addressed several implementation issues to make it practical. For example, they added multiplexers in between the shift register and product lines so that single flip-flops may be connected to product lines. They managed to compact the layout and fit in flip-flops within two pitches of product lines. Furthermore, they proposed a modification of the input decoder so that a universal test pattern may be generated.

Upadhaya and Saluja proposed [33] a BIST PLA architecture as an extension of work by Saluja et al. [27,28]. The scheme is capable of detecting single and multiple SA faults as well as single and multiple crosspoint faults. Fig. 6.10 illustrates the scheme. The augmented PLA, besides the bare PLA included two shift registers TPG1 and TPG2 of $(2n+1)$ and m bits, respectively. These registers apply test patterns. An extra row, Z1, having a crosspoint switch with every column of the PLA is added to the OR plane. Similarly, a column, Z2, having a crosspoint switch with each output line is added. Furthermore, a MUX circuit, and an evaluator circuit, RE, consisting of two counters and a comparator are also added. The counter C1 is for counting the number of 1's generated in the Z1 and Z2 lines. The crosspoints in a column in the AND plane are counted using the Z1 line and crosspoints in the same column in the OR plane are counted using the Z2 line.The counter C2 is of the same length as C1 and contains the number of designed crosspoint switches in the PLA. At the end of the test, values of C1 and C2 are compared. A fault is said to be detected if these values differ from each other.This scheme has a test length of $m(2n+l)$ vectors where $2n$ is the number of inputs, m is the number of columns, and l is the number of outputs. The test vectors are independent of the implemented function which makes it attractive for BIST application. However, a disadvantage of the scheme is its large hardware overhead. Large number of flip-flops is needed to implement shift registers and counters. Besides, an extra row, an extra columns, and a MUX are needed.

6.6 Detection of Realistic Faults in PLAs

The stuck-at fault model is inadequate for representing the wide classes of defects that are possible in a PLA. Tamir and Sequin [30] developed a PLA fault model based upon realistic physical defects. They argued that a missing cp fault in the AND array has the same effect as a weak logic 1 on an input line such that instead of turning on the crosspoint transistor is turned off. Similarly, a missing cp fault in the OR array is equivalent to a weak logic high value on the corresponding product line. A break in an input line can cause the line to float which is equivalent to a weak 0 and/or weak 1 fault. Therefore, if weak 0/

1 faults on input lines are considered, then there is no need to consider the break faults on the input lines. Furthermore, they suggested that troublesome sequential faults are not possible on product and output lines since these lines are either connected to pull up transistors and a break will cause the line to have a stuck-at-0 behavior, or a short to drains of crosspoint transistors. A break in them is the same as a missing/extra crosspoint fault.

Fujiwara [11], besides considering conventional faults (i.e., SA and cp faults), also considered bridging faults. He proposed two new augmented PLA architectures for universal testability that solved the problems of extra hardware and performance degradation associated with previous solutions. The architecture has following distinctive features:

- **Universal Test:** The augmented PLA can be tested with function independent test patterns. Under the single fault assumption, both input test patterns and output responses are function independent. However, under multiple fault assumption the outputs test responses are not function independent.

- **Reduced Overhead:** The amount of extra hardware is substantially reduced compared to previous designs of universally testable PLAs.

- **High Fault Coverage:** The universal test ensured a very high fault coverage not only for SA and cp faults but also for adjacent line bridging faults in the PLA.

- **BIST Application:** The application of BIST is simple since under the single fault assumption, test patterns and output response are function independent.

The first augmented PLA is illustrated in Fig. 6.11. The extra hardware includes, a shift register, a control array (C), an extra column in the AND array, and an extra row in the OR array. The shift register selects a desired column (product line) in the AND array. An arbitrary column is selected by setting logic 0 in the corresponding flip-flop of the shift register and by setting logic 1 in the rest of the shift register. Similarly, the control array in combination with input data can be used to select a given row of the AND array, i.e., to sensitize

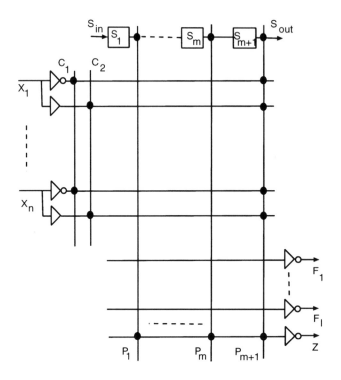

Fig. 6.11: Augmented PLA by Fujiwara for multiple fault detection.

any output line of the decoder. The extra column in the AND array is used to test stuck-at, crosspoint, and bridging faults in the decoder and the control array. It is also used for testing stuck-at and bridging faults on the rows of the AND array. The extra row and output Z of the OR array can be used to test stuck-at and bridging faults on the columns and product lines and crosspoint faults in the AND array. Multiple faults could be detected in this PLA, however, the outputs of the PLA are functions of the implemented logic. The second augmented PLA is similar to the first one except for an extra column in the AND array and an extra row in the OR array. In this configuration, the output response of the PLA under single fault assumption is universal as well.

The augmented PLA proposed by Fujiwara (Fig. 6.11) was extended by Fujiwara et al. [10,31] for BIST application. The overhead of such a scheme is

marginally higher than that of proposed by Fujiwara [11]. The scheme has two
extra columns in the AND plane and one extra row in the OR plane. The output response is compressed into a parity bit which is exclusive-ORed with a
bit representing the cumulative parity of all previous output responses to obtain the new cumulative parity bit. The testing scheme was function independent and had $2m(n+1)+1$ test vectors. The fault coverage of the scheme
included all single and almost all multiple crosspoint faults, SA faults, and
bridging faults.

Liu and McCluskey proposed a BIST scheme for large embedded PLAs [19]
shown in Fig. 6.12. Their fault model included SA faults, bridging faults,
crosspoint faults and switch level faults for dynamic PLAs. The dynamic PLA
is attractive for embedded applications owing to smaller area, speed of operation, etc. An example of dynamic PLA is depicted in Fig. 6.13. The analyzed
dynamic PLA by Liu and McCluskey was a bit different than that shown in
Fig. 6.13. It had a common pull-down transistor to evaluate all product lines in
a plane. In a dynamic PLA, there are three classes of devices: crosspoint transistors, pull-down, and pull-up devices. A stuck open fault in a crosspoint
transistor is equivalent to a missing crosspoint fault. A stuck open fault in a
pull-down transistor is equivalent to multiple SA1 faults on the product lines.
A stuck open fault needs two vectors for test. In this case, the first vector is
provided by the precharge phase. The second vector is provided by the inputs
for testing the multiple product line SA1 fault. A stuck open fault in a pull-up
transistor is not equivalent to any conventional fault. For detection of this
fault, the product line should first be initialized to 0 and then charged to 1
through the faulty transistor.

Transistor stuck on fault detection is quite different from the detection of transistor stuck open faults. A stuck on fault in a crosspoint transistor is equivalent
to extra crosspoint fault. A stuck on fault on the pull-down transistor will result in incomplete precharge (not up to VDD) of product lines depending upon
the transconductance of pull-up and pull-down transistors, respectively. Normally, P-dominant PLA design is recommended to enhance the stuck on testability. The P-dominant design means that the transconductance of every P-

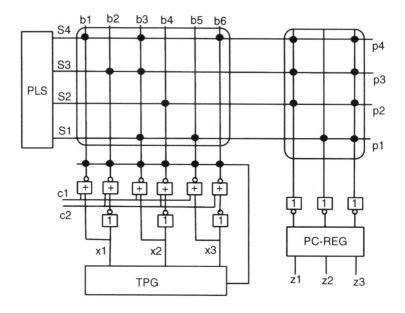

Fig. 6.12: A PLA BIST scheme by Liu and McCluskey.

channel transistor is higher than that of every N-channel transistor. Now, a stuck on fault in a pull-up device in P-dominant device will cause the particular product line to have a detectable SA1 behavior. However, the same fault in N-dominant PLA is not detectable.

The augmented BIST PLA proposed by Liu and McCluskey is illustrated in Fig. 6.12. It has following additional circuitry:

- A test pattern generator (TPG) capable of producing walking 1 as well as walking 0 patterns.

- Modified input decoders allow independent control of true and complement bit lines through control lines c_1 and c_2.

- Two extra product lines are added to the AND plane. First product line has crosspoint connections with every bit line and its output is connected to the TPG. The other extra product line is connected to each bit line in the AND array and each output line in the OR array in such a fashion that the number of crosspoint connections on these lines is odd.

- A product line selector (PLS) provides the ability to activate one product line at a time.

- An extra output line that realizes the logical OR of all product lines is added to the OR array.

- A parity checking register (PC-REG) is connected to the output of the PLA. It functions as a parity counter and a shift register. The PC-REG computes the parity data generated by the test patterns and its contents are shifted out for comparison after the test.

- The BIST scheme proposed Liu and McCluskey requires $2m(n+1)+1$ test vectors, where n is the number of inputs and m is the number of product lines. Furthermore, the augmented PLA contains $(m/4+n+p)$ flip-flops and $(m+n)$ logic gates.

As mentioned before, bridging fault is the dominant fault type in modern CMOS processes. Bridging faults in PLAs are more likely than in random logic owing to the extensive, closely spaced interconnects. Chandramouli et al. [6] analyzed bridging faults in a CMOS PLA and observed that either bridging faults in previous works were ignored [2,8,13,22,23,29] or a wired AND behavior was assumed [11,30]. This assumption is not valid in MOS technologies. Furthermore, PLAs implemented with dynamic logic put extra constraints on bridging fault detection owing to dynamic nature of the PLA. Dynamic PLAs are preferred due to their smaller area, lower power dissipation, and the ability to pipeline the processing for increased throughput. Typically they are implemented as INV-NOR-NOR-INV structures. Fig. 6.13 illustrates an example of a dynamic PLA configuration with 3 inputs, 4 product terms, and 3 outputs.

The distinctive feature of a dynamic PLA is precharge and evaluation lines in AND and OR planes. A two phase non-overlapping clock scheme is generally used. During ϕ_1 phase the AND plane is evaluated and the OR plane is precharged. During the ϕ_2 phase, OR plane is evaluated and AND plane is pre-

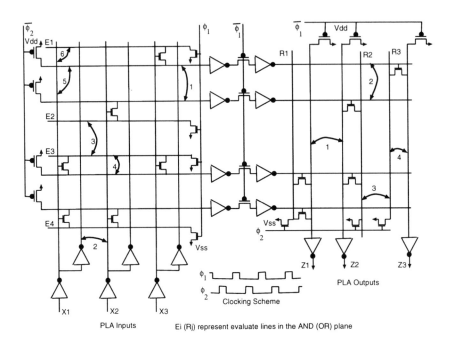

Fig. 6.13: A dynamic PLA with possible bridging faults.

charged. Dynamic latches are placed between the AND and OR planes to buffer the AND plane output. The precharge and evaluate lines in a dynamic PLA are usually rearranged during layout to make efficient use of the area. Therefore, in AND plane sometimes two product lines are adjacent to each other and sometimes two evaluation lines are adjacent. Similarly, in the OR plane sometimes sum lines are adjacent and sometimes two evaluation lines are adjacent [6]. Four basic types of bridging faults are possible in AND and OR planes as illustrated in Fig. 6.13.

- **Type 1:** These are bridging faults between product lines in the AND plane and between sum lines in the OR plane. These faults are similar to the output of a dynamic logic gate that is precharged by a PMOS transistor and evaluation is carried out by NMOS transistors. Such faults are only detected by functional testing. They are not detected by I_{DDQ} since the voltage conflict across the fault can not be sustained in steady state.

- **Type 2:** These are bridging faults between input lines in the AND plane and product lines in the OR plane. These faults are similar to the bridging faults in completely static CMOS circuits. The I_{DDQ} should be used to detect these faults. The detection of such faults by functional testing is not efficient without a DfT scheme. Such a scheme may cost extra area and reduced performance. To excite a bridging fault between any two input lines in the AND plane a single test vector is needed. The requirement for this vector is that all adjacent bit lines in the AND plane as well as all adjacent product lines in the OR plane should have complementary logic values. Therefore, voltage across a bridging fault results in elevated I_{DDQ}.

- **Type 3:** Bridging faults between adjacent evaluate lines, e.g., E2 and E3, (defect 3) are put under this category. These faults do not affect the function of the PLA and are not detectable.

- **Type 4:** This category contains the defects between precharge and evaluate lines (defect 4). These faults are identical to crosspoint device stuck-on faults. These can be detected by test strategies for cp faults. These faults are not detected by I_{DDQ} since both, precharge and evaluation, lines are dynamically excited (i.e., precharged logic high and evaluated low). Therefore, a logical conflict can not be created and sustained for I_{DDQ} testing.

Besides, above mentioned four categories, there are other bridging faults possible (faults 5 and 6). These faults are between input lines and product (or evaluation) lines. Such faults are less likely since affected lines are routed on different metalization levels.

6.7 I_{DDQ} Testable Dynamic PLAs

Bridging defects in dynamic PLAs are more difficult to detect with I_{DDQ} compared to static PLAs. The above mentioned study by Chandramouli et al. [6] confirms this hypothesis. Only type 2 bridging faults are detected by I_{DDQ} testing. Furthermore, transistor leakage faults in a dynamic PLA are not detected by I_{DDQ} based test. However, bridging as well as leakage faults can be

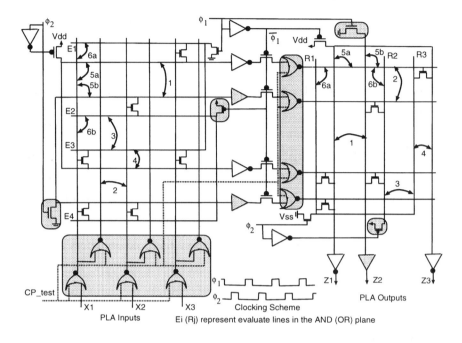

Fig. 6.14: An I_{DDQ} testable dynamic PLA.

detected with I_{DDQ} testing if minor PLA architectural modifications are undertaken. In this section, a DfT methodology for bridging and leakage fault detection in dynamic PLAs by I_{DDQ} is suggested. Furthermore, those faults (e.g., missing cp faults and opens) that are not detected by I_{DDQ} test may be detected by test and DfT schemes illustrated in previous sections. The methodology may be applied with some modifications to a static PLA as well. In the previous chapter, an I_{DDQ} based RAM parallel test approach was illustrated. Similar to RAMs, PLAs also have a very regular architecture where application of I_{DDQ} may result in substantial test cost reduction provided the architecture is modified to facilitate I_{DDQ} testing [26]. Besides the reduction in test cost, the test is independent of the function implemented by the PLA.

6.7.1 First Configuration

Fig. 6.14 illustrates the schematic of a dynamic PLA with DfT features. The figure is same as Fig. 6.13 except for highlighted areas. In a dynamic PLA, the corresponding precharge (product lines in AND plane or sum lines in OR plane) and evaluation lines should be complementary. For example, if the product lines are precharged to VDD, then evaluation should be done to VSS (Fig. 6.13) or vice-versa. So long as precharge and evaluation to complementary logic levels constraint is met, one has the freedom to choose the VDD or VSS for individual product or sum lines for precharge (evaluation). This feature has been exploited to enhance the I_{DDQ} testability for PLAs. The odd product lines in the AND plane and the odd sum lines in the OR plane are precharged to VDD as before. However, the even product lines in the AND plane and the even sum lines in the OR plane are precharged to VSS (shaded NMOS transistors). Similarly, corresponding even evaluation lines in the AND and OR planes are also modified. These lines are evaluated to VDD (shaded PMOS transistors). The output of even product and sum lines is buffered with non-inverting buffers so as to maintain the proper logic operation. A test control signal CP_test is provided. The CP_test in normal mode is kept at logic low that ensures the normal PLA operation. At the instance when CP_test is high, all bitlines in the AND plane and all sum lines in the OR plane are pulled down ensuring no crosspoint transistor is on.

6.7.1.1 Bridging Fault Detection

Let us consider, the above mentioned four categories of bridging faults once again. For the first I_{DDQ} measurement, both clock phases are kept high. This will ensure that all precharge lines and all evaluation lines in both planes are active at the same time. Now, depending upon input stimuli conditions, some crosspoint transistors may be on, which will invalidate the test. Therefore, CP_test is also kept high so that all crosspoint transistors are in off state. Since, all adjacent precharge and evaluation lines are driven to complementary logic levels, any bridging fault among them will result into elevated quiescent current. Therefore, all types 1,3 and 4 faults as well as leakage faults in all crosspoint transistors (e.g., stuck-on) in both the planes will be detected by this measurement. This test is independent of the function implemented in the PLA.

Table 6.1: Bridging and leakage faults and their detection conditions for I_{DDQ} testable PLA of Fig. 6.14.

Test	Test Conditions	Detected Faults	Comments
I1	$\Phi1=1, \Phi2=1,$ CP_test=1,	AND pl: typ 1,3,4, all stuck-on cp faults; OR pl: typ 1,3,4, all stuck-on cp faults	Function indepen. test
I2	$\Phi1=1, \Phi2=0,$ CP_test=0 Inputs with compl. data	AND pl: type 2, 3 faults	Function indepen. test
I3	$\Phi1=1, \Phi2=0,$ CP_test=0 Inputs with appropriate data	OR pl: type 2 faults	Function depen. test

The second I_{DDQ} measurement is needed to test for type 2 faults in the AND plane. During this measurement, the clock phase ϕ_1 is kept high and the clock phase ϕ_2 is kept low, and the test signal CP_test is kept low. These conditions ensure that no crosspoint transistor is on in the AND plane. Now, adjacent inputs are driven to complementary logic levels such that all type 2 faults are activated in the AND plane and are detected by elevated quiescent current. This test is also independent of the function implemented by the PLA. Similarly, type 2 faults in the OR plane can also be detected by keeping adjacent product lines to complementary logic values. However, the test effectiveness is dependent of the implemented function in the AND plane. Table 6.1 shows these three I_{DDQ} measurement with test conditions and detected faults by each measurement.

Bridging faults 5, 6 and their derivative are detectable without a DfT scheme. Here, we need to distinguish a bridge depending upon whether it is affecting an odd or even product (or evaluation) line. This is because alternative product (or evaluation) lines are precharged (evaluated) to complementary logic levels requiring different fault detection conditions. However, in general, these defects are detected by a proper combination of input data, clock phases, and CP_test. For example, logic high on both clock phases together with high

CP_test will ensure detection of faults 5a and 6b in both the planes. Similarly, other faults can also be detected.

6.7.1.2 Discussion

These three I_{DDQ} measurements will not detect open defects. An open defect in a precharge transistor will cause the corresponding product or sum lines to have multiple SA0 or SA1 faults. Similarly, an open defect in evaluation lines will also cause multiple SA0 or SA1 faults that are easy to detect by logic testing. An open defect may cause a crosspoint transistor to be open, which needs to be tested either by functional testing or by structural testing with DfT schemes mentioned in previous sections.

In this DfT scheme, even product lines are evaluated to logic high through NMOS crosspoint transistors. Such a scheme, will result in the evaluation to $VDD - V_{Tp}$ voltage on even product lines. Therefore care should be taken to size the subsequent buffer such that it takes into account the threshold voltage drop. The threshold voltage drop on even product lines may be an issue for robust design. It may also result in DC power dissipation in the buffer, resulting in increased power consumption. Finally, in low voltage application, it may result in unacceptably reduced noise margins.

There are several possible solutions to the problem of threshold voltage drop on the even product and sum lines. Replacing NMOS crosspoint transistors with PMOS crosspoint transistors for even product and sum lines is probably the simplest. Such an arrangement will not cause the threshold voltage drop. For example, testing of type 4 bridging defects will require an elaborate arrangement since forcing logic 0 will not switch off PMOS crosspoint transistors. Furthermore, there will be constraints on the logic implementation of AND and OR planes. Application of latches or sense amplifiers instead of inverters (buffers) to restore the logic level is yet another solution. However, it may increase the complexity of the PLA and reduce performance without really adding a significant benefit.

6.7.2 Second Configuration

In situations where the threshold voltage drop on the even product lines is unacceptable, the previous DfT scheme may be suitably modified. Fig. 6.15 illustrates a modified version of the previous DfT scheme that retains the fault coverage of the previous scheme, alleviates the problem of the threshold voltage drop, and does not result in excessive area and performance penalties. However, this scheme requires two additional test control signals. Signal Br_test is provided to control even product and evaluation lines in AND and OR planes in the test mode. Similarly, signal OR_test is provided to control evaluation lines in the OR plane in the test mode. In the normal mode, Br_test as well as OR_test signals are kept at logic low and the PLA functions as usual, i.e., product and sum lines are precharged to logic high and evaluated to logic low. However, under test conditions these signals are driven logic high to excite different types of faults. These conditions will be explained in subsequent paragraphs. It is worthwhile to note that there may be situations where it is not possible to have a total of three inputs for test purposes. In such cases, an on-board state machine can be designed with only one or two inputs whose outputs may be decoded to derive these signals. Alternatively, CP_test may be decoded from the two clock phases. As it is clear from Tables 6.1 and 6.2, a high CP_test is needed only when both clock phases are high. Therefore, this signal may be derived from clock phases.

6.7.2.1 Bridging Fault Detection

Detection of bridging faults in this PLA is similar to the previous one. For type 1, 3, and 4 bridging faults in the AND plane following conditions are required. CP_test is kept at logic high so that all crosspoint transistors in the PLA are in non-conducting state. The clock phase ϕ_1 is kept high which drives logic low on odd evaluation lines. The clock phase ϕ_2 is also kept high which enables odd product lines to be logic high. Furthermore, Br_test is kept at logic high which enables even product lines to be driven logic low and even evaluation lines to be driven logic high in AND and OR planes, respectively. Under these conditions, any bridging fault of above mentioned categories in

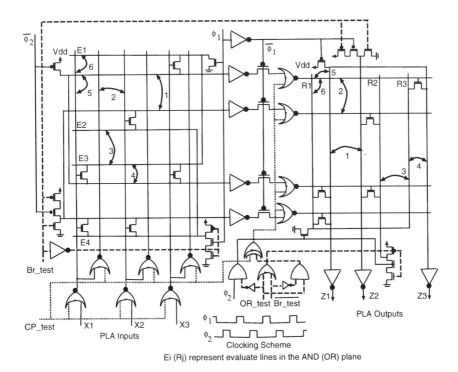

Fig. 6.15: An alternative DfT scheme for I_{DDQ} testable PLAs.

both planes as well as leakage faults in all crosspoint transistors in both the planes will be detected by this measurement. This measurement is independent of the function implemented in the PLA. This measurement is shown as the first measurement in Table 6.2.

The detection of type 2 bridging faults in the AND plane requires an additional I_{DDQ} measurement. In this measurement, adjacent input lines in the layout are driven to complementary logic values. The clock phase ϕ_1 is kept high and the clock phase ϕ_2 is kept at logic low. Furthermore, all test inputs (CP_test, Br_test, and OR_test) are also kept at logic low. In other words, the PLA is kept in the normal mode and the fault is detected by giving appropriate inputs. Such an arrangement will ensure the detection of bridging faults between input lines. This measurement is shown as the second measurement in Table 6.2 and is independent of the function implemented in the PLA.

Table 6.2: Bridging and leakage faults and their detection conditions for I_{DDQ} testable PLA of Fig. 6.15.

Test	Test Conditions	Detected Faults	Comments
I1	$\Phi1=1, \Phi2=1$, CP_test=1, Br_test=1, OR_test=0	AND pl: typ 1,3,4, all stuck-on cp faults; OR pl: typ 1,3,4, all stuck-on cp faults	Function indepen. test
I2	$\Phi1=1, \Phi2=0$, CP_test=0 Br_test=0, OR_test=0 Inputs with compl. data	AND pl: type 2, 3 faults	Function indepen. test
I3	$\Phi1=1, \Phi2=1$, CP_test=1 Br_test=1, OR_test=1	OR pl: type 1, 2 faults	Function indepen. test

Detection of type 2 faults in the OR plane requires explanation. In order to detect these faults independent of the implemented function the following scheme is applied. The clock phase ϕ_2 is kept high, which enables odd product lines to be logic high. Br_test is also kept high, which ensures that even product lines are driven logic low. CP_test is kept high so that all crosspoint transistors in AND plane are in off state and do not invalidate the test.

Now, the clock phase ϕ_1 is kept high which enables the product lines to drive the OR plane. However, for that to happen CP_test should not be high because a high CP_test will make all the product lines in the OR plane low. In such a condition type 2 defects will not be detected. Therefore, CP_test is EXORed with logic AND of Br_test and OR_test. These two signals should be high for the detection of type 2 bridging faults in OR plane. Essentially, when both signals are high the application of CP_test on OR plane is disabled. Such an arrangement ensures that the adjacent product lines in the OR plane are driven to complementary logic values. Unfortunately, keeping both clock phases high excites sum and evaluation lines to complementary logic values that may cause leakage through crosspoint transistors depending upon the logic state of the product lines. Therefore, to avoid invalidation of the test, an addition signal, OR_test, is provided to control the evaluation lines in the high impedance state. The OR_test signal which is normally kept at logic low, is kept at logic high for this test. Under these conditions a type 2 bridging fault will give rise to an elevated I_{DDQ} level.

Detection of bridging faults 5 and 6 is simpler in this configuration since all product lines are precharged high and all evaluation lines are evaluated low. However, similar to the previous configuration a proper combination of input data, clock phases, and CP_test ensures the detection of these faults. For example, logic high on both clock phases together with high CP_test will ensure detection of fault 5 in both planes. Similarly, other faults can also be detected.

6.8 Conclusion

Programmable logic circuits are widely used in VLSI circuits and systems since they provide a simple method of implementing complex functions. These devices range from simple programmable logic arrays (PLAs) to high complexity field programmable gate arrays (FPGAs) and complex programmable logic devices (CPLDs). In its simplest form a PLA is a highly uniform structure capable of implementing any Boolean function expressed as a sum of products. PLAs at chip level are commonly used to design instruction decoders of the microprocessors, and combinational circuitry of finite-state machines. Their popularity comes from the fact that they are simple to design, owing to their regular architecture and the availability of automated synthesis programs.

At the board or system level, user programmability offers flexibility of quick prototyping, last minute configurations, etc. Therefore, these devices are sold as dedicated, user programmable chips. However, two levels of configurable architecture gives rise to poor speed performance and expensive manufacturing costs. Therefore, the PLA architecture gave way to PAL architecture with only a single level of programmability, a wired AND-plane, and fixed OR gates. Single level of programmability results in some constraints in arbitrary logic implementations. However, plurality of logic gates in fixed OR-plane are usually sufficient for high utilization. Furthermore, PAL devices come in a number of configurations, with different numbers and different sizes of OR gates. In order to implement sequential circuits, usually flip-flops (register) are added on the inputs and outputs of a PAL.

The complexity of modern PLDs comes through programmable architecture that includes logic, SRAM, EEPROM, and a large number of programmable interconnects on the same substrate. Such complex devices are difficult to model with conventional fault modeling techniques. Another distinctive feature of modern PLDs is the re-programmability which adds to test complexity. Early PLA and PAL devices were one time programmable (OTP) devices. These structures were either personalized at fab or by the end-user. Therefore, a test for such devices was often limited by the implemented functions. However, re-programmability necessitates a test for total functionality including the programmable devices. There have been efforts to determine tests that are independent of the function implemented (universal tests) in the PLA. In order to appreciate the test complexity of modern programmable logic devices, it is important to review the basic PLA test issues.

In this chapter we focused on PLAs because PLAs are still popular with embedded applications compared to the PAL architecture. Moreover, the test complexity of PAL devices with only one programmable plane is lower than that of PLAs. Furthermore, in this book, we are primarily concerned with testing of defects in embedded building blocks. Finally, the basics of PLA testing have relevance in the testing of present day CPLD and FPGA architectures as well.

References

1. V.K. Agarwal, "Easily Testable PLA Design," Chapter 3 in *VLSI Testing*, Edited by T.W. Williams, Elsevier Science Publishing Company, Inc., pp. 65-93, 1986.

2. V.K. Agarwal, "Mutiple Fault Detection in Programmable Logic Arrays," *IEEE Transactions on Computers,* vol. C-29, pp. 518-522, June 1980.

3. S. Bozorgui-Nesbat and E.J. McCluskey, "Lower Overhead Design for Testability for Programmable Logic Arrays," *Proceedings of the International Test Conference,* 1984, pp. 856-865.

4. S.D. Brown, *Field-Programmable Devices: Technology, Applications, Tools,* Los Gatos: Stan Baker Associates, 1995.

5. S. Brown and J. Rose, "FPGA and CPLD Architectures: A Tutorial," *IEEE Design & Test of Computers,* vol. 12, pp. 42-57, Summer 1996.

6. V. Chandramouli, R. Gulati, R. Dandapani and D.K. Goel, "Bridging Faults and Their Implication to PLAs," *Proceedings of International Test Conference,* 1990, pp. 842-859.

7. W. Daehn and J. Mucha, "A Hardware Approach to Self Testing of Large Programmable Logic Array," *IEEE Transaction on Computers,* vol. C-30, pp. 829-833, November 1981.

8. H. Fujiwara, K. Kinoshita and H. Ozaki, "Universal Test Sets for Programmable Logic Arrays," *Digest of 10th International Syposium on Fault Tolerant Computing,* 1980, pp. 137-142.

9. H. Fujiwara and K. Kinoshita, "A Design of Programmable Logic Arrays with Universal Tests," *IEEE Transactions on Computers,* vol. C-30, pp. 823-828, 1981.

10. H. Fujiwara, R. Treuer and V.K. Agarwal, "A low Overhead, High Coverage, Built-In Self-test PLA design," *Digest of 15th International Syposium on Fault Tolerant Computing,* 1985, pp. 112-117.

11. H. Fujiwara, "A New PLA Design for Universal Testability," *IEEE Transactions on Computers,* vol. C-33, pp. 745-750, 1984.

12. P.P. Gelsinger, "Built In Self Test of The 80386," *Proceedings of International Conference on Computer Design,* 1986, pp. 169-173.

13. S.Z. Hassan and E.J. McCluskey, "Testing of PLAs Using Multiple Parallel Signature Analyzers," *Digest of 13th International Symposium on Fault Tolerant Computing,* 1983, pp. 422-425.

14. S.J. Hong and D.L. Osatpko, "FITPLA: A Programmable Logic Array for Function Independent Testing," *Digest of 10th International Syposium on Fault Tolerant Computing,* 1980, pp. 131-136.

15. K.A. Hua, J.-Y. Jou and J.A. Abraham, "Built In Tests for VLSI Finite State machines," *Digest of 14th International Symposium on Fault Tolerant Computing,* 1984, pp. 422-425.

16. J. Khakbaz, "A Testable PLA Design with Low Overhead and High Fault Coverage," *Digest of 13th International Syposium on Fault Tolerant Computing,* 1983, pp. 426-429.

17. J. Khakbaz, "A Testable PLA Design with Low Overhead and High Fault Coverage," *IEEE Transactions on Computers,* vol. C-33, pp. 743-745, August 1984.

18. J. Kuban and J. Salick, "Testability Features of the MC68020," *Proceedings of International Test Conference,* 1984, pp. 821-826.

19. D.L. Liu and E.J. McCluskey, "Design of Large Embedded CMOS PLA's for Built-In Self-Test," *IEEE Transactions on Computer Aided Design,* vol. 7, no. 1, pp. 50-59, January 1988.

20. Y. Min, "A PLA Design for Ease of Test Generation," *Digest of 14th International Syposium on Fault Tolerant Computing,* 1984, pp. 436-442.

21. E.I. Muehldorf, G.P. Papp and T.W. Williams, "Efficient Test Pattern Generation for Embedded PLAs," *Proceedings of the International Test Conference,* 1980, pp. 349-358.

22. D.L. Osatpko and S.J. Hong, "Fault Analysis and Test Generation for Programmable Logic Arrays," *IEEE Transactions on Computers,* vol. C-28, pp. 617-626, September 1979.

23. Rajaski and Tyszer, "The Influence of Masking Phenomenon on Coverage Capability of Single fault Test Sets in PLAs," *IEEE Transactions on Computers,* vol. C-35, no. 1, pp. 81-85, January 1986.

24. R. Rajsuman, *Digital hardware Testing: Transistor-Level Fault Modeling and Testing,* Boston: Artech House, Inc. 1992.

25. K.S. Ramanatha and N.N. Biswas, "A Design for Testability of Undetectable Crosspoint Faults in PLAs," *IEEE Transactions on Computers*, vol. C-32, no. 6, pp. 551-557, June 1983.

26. M. Sachdev, "I_{DDQ} Testable Programmable Logic Arrays," *European Patent Application no. 97200847.8*, March 1997.

27. K.K. Saluja, K. Kinoshita and H. Fujiwara, "A Multiple Fault Testable Design of Programmable Logic Arrays," *Digest of 11th International Syposium on Fault Tolerant Computing*, 1981, pp. 44-46.

28. K.K. Saluja, K. Kinoshita and H. Fujiwara, "An Easily Testable Design of Programmable Logic Arrays for Multiple Faults," *IEEE Transactions on Computers*, vol. C-32, pp. 1038-1046, 1983.

29. J. Smith, "Detection of Faults in Programmable Logic Arrays," *IEEE Transactions on Computers*, vol. C-28, pp. 845-853, November 1979.

30. Y. Tamir and C.H. Sequin, "Design and Application of Self-Testing Comparators Implemented with MOS PLAs," *IEEE Transactions on Computers*, vol. C-33, no. 6, pp. 493-506, June 1984.

31. R. Treuer, H. Fujiwara, and V.K. Agarwal, "Implementing a Built-In Self Test PLA Design," *IEEE Design and Test of Computers magazine*, vol. 2, pp. 37-48, April 1985.

32. S.M. Trimberger, *Field Programmable Gate Arrays*, Boston: Kluwer Academic Publishers, 1994.

33. S.J. Upadhyaya and K.K. Saluja, "A New Approach to the Design of Built-In Self Testing PLA's for High Fault Coverage," *IEEE Transactions on Computer Aided Design*, vol. 7, no. 1, pp. 60-67, January 1988.

34. . S. Yajima and T. Aramaki, "Autonomously Testable Programmable Logic Arrays," *Digest of 11th International Syposium on Fault Tolerant Computing*, 1981, pp. 41-43.

CHAPTER 7

Defect Oriented Analog Testing

Analog circuits due to their non-binary operation are influenced by process defects in a different manner compared to digital circuits. Seemingly an innocuous defect for digital logic may cause unacceptable degradation in analog circuit performance. This chapter surveys the advances in the field of defect oriented analog testing and summarizes strengths and weaknesses of the method for analog circuits.

7.1 Introduction

In the previous chapters we demonstrated the application of defect oriented test techniques on solving digital and quasi-digital (RAM) test problems with reasonable success. In this chapter we apply the same methodology to analog circuits. However, analog test complexity is different from that of digital circuits. The emergence of mixed signal ICs further complicates the test issues. In general, analog testing poses challenges still to be surmounted by researchers. Several reasons are attributed to the inherent analog test complexity [7,9,41,53] and a number of solutions have been suggested [4,7-9,12-15,24-27,30,31,38,40,41,48,50-53]. However, in spite of these attempts and proposed solutions, almost all analog circuits are presently tested in a functional manner.

For any test strategy to succeed in terms of test quality and global applicability, it should have a sound basis. For example, poor performance of the stuck-at model based digital test schemes amply demonstrate how without a firm basis, test strategies can fail to deliver quality products [21,43]. Therefore, we set the following objectives:

- To propose an analog test methodology based on a firm foundation. The proposed test strategy is based on manufacturing process defects that provide an objective basis for analog fault model development and test generation.

- To assess the effectiveness of the methodology from two standpoints: (a) contribution of inductive fault analysis (IFA) towards testing silicon devices in the production environment and, (b) contribution of IFA towards robust analog design against process defects, quantifying the fault coverage of analog tests, and examining the practicality of analog DfT schemes.

Analog circuits, due to their non-binary circuit operation, are influenced by defects in a different manner compared to digital circuits. This poses additional challenges for modeling of defects in analog circuits. In fact, the analog fault modeling is identified as a critical factor in the success of any analog DfT scheme [45]. Furthermore, we explore the concept of structural test vectors in analog domain and examine the potential of simple test stimuli in fault detection.

7.2 Analog Test Complexity

Considerable effort has been devoted to identify the causes of the analog test complexity [7,9,33,41,53]. These are summarized as follows:

- Unlike digital circuits, analog circuits do not have the binary distinction of pass and fail. The time and voltage continuous nature of their operation makes them further susceptible to defects. Therefore, test procedures are needed to discriminate between various faulty conditions and the non-faulty condition.

- Analog systems are often non-linear and their performance heavily depends on circuit parameters. Process variations within allowable limits can also cause unacceptable performance degradation. Deterministic methods for modeling such variations are often inefficient.

- In digital circuits, the relationship between input and output signals is logical (Boolean) in nature. Many digital DfT schemes simplify this relationship to reduce the test complexity. On the other hand, the input-output relationship in analog circuits is non-Boolean. Such behavior is complex and difficult to model.

- Digital DfT schemes based on structural division of the circuit, when applied in analog domain, are also largely unsuccessful because of their impact on the circuit performance.

- In digital domain, there exist a wide range of well defined and industrially accepted fault models. These models or abstractions form the basis for representing the faulty circuit behavior as well as test pattern generation. In analog domain the effectiveness of these models is questionable. Moreover, in the absence of an acceptable fault model, test generation has been ad-hoc and testing has been largely functional (specification oriented) in nature.

- Since different specifications are tested in different manners, analog functional testing is costly and time consuming. Moreover, often extra hardware is needed to test various specifications.

- Limited functional verification does not ensure that the circuit is defect-free and escaped defects pose quality and reliability problems.

Analog testing also suffers from automatic test equipment (ATE) related issues. For example, when noise level in test environment is not acceptable, complex shielding techniques are required. Furthermore, the integrity of test depends on interface, interconnections, probe card, etc.

7.3 Previous Work

Analog fault modeling and diagnosis received much theoretical attention in
the late 1970s and 1980s. Duhamel and Rault presented an excellent review of
the topic [7]. These theoretical works relied on the characteristic matrix of the
circuit under test for testability and diagnosability. Though those methods had
a broad scope, their application to specific circuits has not been successful.
The analog fault detection and classification can broadly be divided into following categories:

7.3.1 Estimation method

This method can further be subdivided into an *analytical* (or deterministic)
method and a *probabilistic* method. In the former, the actual values of the parameters of the device are determined analytically or based on the estimation
criteria (physical or mathematical). The least square criterion approach represents this class. Typically, in this approach a factor of merit, S_i , is associated
with each parameter as:

$$S_i = \sum_{j=1}^{m} \{g_j - y_j(X_i)\}^2 \qquad \text{(EQ 7.1)}$$

where g_j is the measured value of the characteristic y_j , and X_i is a vector x_1,
...., x_n of parameters which have their nominal values, except for x_i. The factor
of merit associated with x_i is taken as the minimum value of S_i . The most
likely faulty parameter is the one that, given all other parameters are at their
nominal value, minimizes the difference between nominal and measured characteristics.

In probabilistic methods the values are inferred from the tolerance of the parameters. For example, inverse probability method is the representative of this
class. Elias [8] applied statistical simulation techniques to select parameters to
be tested. On this basis, he also formulated the test limits.

7.3.2 Topological Method

This method is also known as *simulation-after-test* (SAT) method. The topology of the circuit is known and SAT method essentially reverse engineers a circuit to determine the values of the circuit component parameters. A set of voltage measurements is taken and then numerical analyses determine parameter values [4,12,26,30,31,40,50,53]. SAT methods are very efficient for soft-fault diagnosis because soft faults are based on a linearized network model. However, this method is computation intensive and for large circuits the algorithms can be inefficient.

One of the first theoretical studies of the analog circuit fault-analysis problem was initiated by Berkowitz [4]. He mathematically defined the concept of network-element-value solvability and studied the measurement conditions required to solve the problem. Trick et al. [50] and Navid and Willson Jr. [26] proposed necessary and sufficient conditions for network solvability problem. Trick et al. used only voltage and single frequency sinusoidal input to determine the parameter value for linear passive circuits. Navid and Willson Jr. suggested that for small signal analysis, non-linear active elements, like transistors and diodes, can be linearized around their operating points. They proposed an algorithm covering the solvability problem for linear passive network as well as active devices. Rapisarda and Decarlo [31] proposed the *tableau approach* for analog fault diagnosis instead of transfer function oriented algorithms. They argued that *tableau approach* with *multi-frequency excitation* would provide simpler diagnostic solution. Salama et al. [40] proposed that large analog circuits can be broken into smaller uncoupled networks by nodal decomposition method. These subnetworks can be tested independently or in parallel. Every subnetwork is associated with a *logical variable* σ, which takes the value 1 if the subnetwork is good and 0 if it is faulty. Furthermore, every test is associated with a *logical test function* (LTF) that is equal to the complete product of variables σ_{ji}. If the network passes the test, then

$$T_{J_i} \equiv \sigma_{j1} \cap \sigma_{j2} \cdots \cap \sigma_{jk} \qquad \text{(EQ 7.2)}$$

where

$$I_t \equiv j_1, j_2, ..., j_k \qquad\qquad \text{(EQ 7.3)}$$

j_i refers to network S_{j_i} , k is the number of subnetworks involved in the test.

Hemink et al. [12] postulated that the solvability of the matrix depends on the determination accuracies of the parameter. The set of equations describing the relations between parameters and measurements can be *ill-conditioned* due to 'almost' inseparable parameters. Further, they contended, that solving such a set of equations inevitably leads to large computation errors. They overcome this problem by an improved algorithm that finds the sets of separable high-level parameters and computes the determination accuracy of the parameters. Recently Walker et al. [53] developed a two-stage SAT fault-diagnosis techniques based on bias modulation. The first stage, which diagnoses and isolates faulty network nodes, resembles the node fault location method. The second stage, a subnetwork branch diagnosis, extracts faulty network parameters. The branch diagnosis is achieved by *element modulation*, a technique that varies the value of the element externally as a modulated element. The diagnostic technique requires a single test frequency and the ability to control the network bias from external source.

7.3.3 Taxonomical method

This method is based upon a fault dictionary. This is also known as *simulation-before-test* (SBT) method [7,9,13,15,24,27,58,41]. The fault dictionary is a collection of potential faulty and the fault-free responses. During the actual test the measured value is compared with the stored response in the dictionary. A fault is detected if at least for a set of measurements the actual response differs from the fault-free response by predetermined criteria. The accuracy of the method depends on the accuracy of the fault dictionary [7].

The fault-free and faulty circuit responses are measured at certain key points. The number of test points depends on the diagnosis resolution and test stimuli. Schemes based on this method can be segregated according to the input stimuli and fault dictionary construction. For example, this method can be implemented with DC signals [13,24], or various time-domain signals [48] or AC signals [27]. The DC fault dictionary approach is simple but it can not detect purely capacitive or inductive defects. Such defects often give rise to parametric or soft faults that are more readily detected by the transient or AC dictionary approach. Slamani et al. [41] made a combined dictionary for DC, transient and AC input stimuli to predict the defective component. They claimed that this method could detect wide ranging defects, from tolerance deviation to catastrophic faults. Sachdev [33] made a similar fault dictionary from the catastrophic processing defect information using inductive fault analysis (IFA) [40].

7.4 Defect Based Realistic Fault Dictionary

Application of a defect oriented approach [40] in solving analog test problems has gained popularity in the recent past [22,33-35,44,45]. It is proposed as an alternative to analog functional testing. However, this proposal is not without controversy. What makes this topic so controversial? The critics of IFA based analog testing are quick to remind that the test issues of analog circuits are more qualitative than quantitative. It is not uncommon to come across an analog circuit having signal to noise ratio (SNR) of 100 dB or operation frequency of few hundred MHz or input-offset voltage less than 20 mV. Secondly, analog circuits often exploit a number of circuit and device level parameters (e.g., transistor matching, layout considerations, transistor sizing, etc.) to achieve the maximum possible performance. Unfortunately, such clever techniques render the circuit vulnerable to several factors since the maximum possible circuit performance is achievable only under the optimal fabrication and operating conditions. Thirdly, in the case of analog circuits, the range of optimal conditions is substantially narrower than that of their digital counterparts. For example, in digital circuits, typically the critical path is the most sensitive for performance (parametric) degradation. While in analog circuits, the parametric requirement is much higher and widely distributed over the circuit layout. Therefore, any sub-optimal performance of one or more parameters may have significant impact on the performance. A good test program should test for all such sub-optimal performances. Finally, one may ask how comprehensive and accurate is the yield loss model based upon defects alone in the case of analog circuits? Since these are formidable concerns, according to critics, the functional (specification) testing is the only alternative to ensure the circuit performance, specifications and quality.

On the other hand, those who have faith in IFA based analog testing will argue that IFA based testing combines the circuit topology and process defect data to arrive at the realistic fault data that is specific to the circuit. This information can be exploited by test professionals to generate effective and economic tests. The same information can be used by analog circuit designers to design robust and defect tolerant circuits. Secondly, this is a structured and globally applicable test methodology that substantially reduces the test generation cost. Finally, they cite numerous examples of digital domain where IFA based tests contributed significantly to test simplification and test quality improvement [40]. We address the assessment of analog IFA from two standpoints: (i) contribution of IFA towards testing silicon devices in the production environment, and (ii) contributions of IFA towards robust analog design against process defects in quantifying the fault coverage of analog tests, and in examining the practicality of analog DfT schemes.

In the classical sense, the defect based fault dictionary can be categorized as a SBT approach. All forms of fault simulation are carried out before the test. Fig 7.1 illustrates basic concepts of the defect based fault dictionary. The manufacturing process defects, catastrophic as well as non-catastrophic, form the core of the methodology. Realistic defects are sprinkled over the circuit to determine the realistic fault classes. These faults are simulated with given test vectors. The fault simulation is carried out in a Spice-like simulator to achieve accurate results. Alternatively, if fault simulation at circuit level is not possible owing to the circuit complexity, a high level model of the circuit may be used. The results of the fault simulation are compiled into a fault dictionary. A fault is considered detected if the faulty response differs from the nominal response by a predetermined criterion. Next, a test program is prepared taking the fault

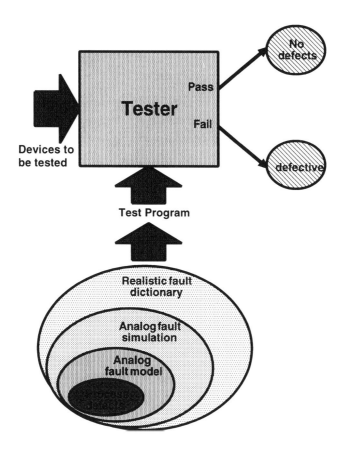

Fig. 7.1: A realistic defect based testability methodology for analog circuits.

dictionary into account. The effectiveness of the fault dictionary depends on several factors. Foremost is the defect population and relative probabilities of occurrence of various defects. It is not possible to carry out an exhaustive fault simulation with all permutations of defect (fault) impedances. Therefore, the effectiveness of the dictionary depends how representative are the faults of actual defects and how accurate is the simulator. Finally, the dictionary effectiveness also depends on pass/fail criterion. Nevertheless, the defect based fault dictionary forms the basis for a structured analog testing.

Mathematically, we can define this concept as follows: let $\mathcal{F}$ be the fault matrix of all faults in a given CUT and let F_0, the first element of the matrix, be the fault-free element. Moreover, let S be the matrix of stimuli applied at CUT inputs and let $\mathcal{D}$ be the matrix of the fault-free and faulty responses (i.e., the fault dictionary). Furthermore, let us assume that in a given circuit, there are n faults, then the size of the fault matrix taking into account the fault-free element as well, is $(n+1).1$. The fault matrix, $\mathcal{F}$, can be written as follows:

$$\mathcal{F} = \begin{bmatrix} F_0 \\ F_1 \\ F_2 \\ \circ \\ \circ \\ F_n \end{bmatrix} \qquad \text{(EQ 7.4)}$$

For the formulation of the stimuli matrix, let us assume that CUT has m inputs. Therefore, any arbitrary test vector S_i consists of s_{i_1}, s_{i_2}, ..., s_{i_m}. In order to simplify the analysis we assume that for any S_i all constituents put together excite the faulty CUT in a particular way. Therefore, the constituents of S_i can be treated as *scalars*. This is not an unreasonable assumption since in analog circuits, unlike digital circuits, the circuit function depends on the continuous operation of its components and stimuli. The analysis will hold even in the absence of this assumption, however, it would require rigorous mathematics. Furthermore, in spite of this assumption, one has total freedom to select the constituents of a given (S_i) stimuli. Hence, the stimuli matrix can be formulated as:

$$S = \begin{bmatrix} S_1 & S_2 & ... & S_i & ... & S_t \end{bmatrix} \qquad \text{(EQ 7.5)}$$

Where t is total number of inputs. The fault dictionary $\mathcal{D}$ is a function of the fault matrix as well as the stimuli matrix.

$$\mathcal{D} = f(\mathcal{F} \times S) \qquad \text{(EQ 7.6)}$$

For each fault detection mechanism such as voltages on different outputs or dynamic current, formulation of different matrices will be required. Alternatively, like the stimuli matrix, different detection mechanisms can be treated as scalar fields of each d_{ij}. Elements of the matrix $\mathcal{D}$ are given as follows:

$$d_{ij} = f_{ij}(F_i \times S_j) \qquad \text{(EQ 7.6)}$$

where $0 \le i \le n$ and $1 \le j \le t$

We simulate the CUT to find out all d_{ij} of the fault dictionary. It is possible to compute these elements when function f_{ij} is known. The first row of $\mathcal{D}$ gives the fault-free responses. The size of $\mathcal{D}$ is $(n+1).t$.

7.4.1 Implementation

The implementation issues of the fault dictionary are segregated as follows: (i) related to defects and fault modeling, and (ii) related to the analysis flow. The former is concerned with collecting of defect data for a given fab, modeling of defects for a given fault simulator, etc. The latter is concerned with establishing an analysis flow, determination of pass/fail criterion, etc.

(i) **Process defects and modeling:** Defects and their impact on the device performance have been studied in detail in the literature [11,18-21,42]. Broadly speaking, causes of IC functional failures can be separated into global and local process disturbances. Global disturbances are primarily caused by defects generated during the manufacturing process. The impact of these global (or manufacturing process related) defects covers a relatively wider chip area. Hence, they are detected before functional (or structural) testing by using simple test-structure measurements or supply current tests. A vast majority of faults that have to be detected during functional (or structural) testing are caused by local defects, popularly known as *spot defects* [18]. Since the global defects are relatively easy to detect by other measurements, we use spot defects for fault modeling purposes. In a typical single poly double metal CMOS process, commonly found spot defects are:

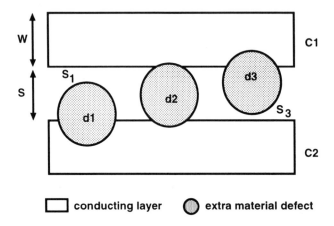

Fig. 7.2: Catastrophic and non-catastrophic shorts.

- Short between wires

- Open in a wire

- Pin hole; oxide, gate oxide, *pn*-junction

- Extra contact or via

- Missing contact or via

Spot defects can also be categorized into two classes: (i) Defects causing a complete short or open in the circuit connectivity. These are often referred as catastrophic defects. (ii) Defects causing an incomplete short or open in the circuit connectivity. These are often called non-catastrophic or soft defects. Fig. 7.2 shows catastrophic and non-catastrophic defects caused due to spot defects between two conducting layers C1 and C2. Defect d2 causes a catastrophic short (low resistance bridge) between both conductors. Therefore, the defect modifies the circuit characteristics and performance drastically. However, defects d1 and d3 do not cause complete shorts but reduce the spacing to S_1 and S_3, respectively. Reduced spacing causes high impedance bridging defects between conductors that can be modeled as a parallel combination of a resistance R and a capacitance C. The values of R and C for d1 are given by following equations:

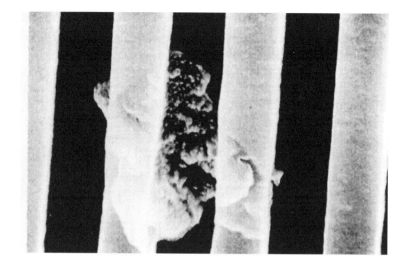

Fig. 7.3: Photograph showing a high resistive short in metalisation layer.

$$R = \frac{\rho_{SiO_2} \times S_1}{A} \qquad \text{(EQ 7.7)}$$

$$C = \frac{\varepsilon_{SiO_2} \times A}{S_1} \qquad \text{(EQ 7.8)}$$

In these equations, ρ_{SiO_2} is the resistivity and ε_{SiO_2} is the permittivity of the insulator between the conductors C1 and C2. The S_1 is the reduced spacing between the conductors which were otherwise a distance S apart. The area between the defect and conductor is represented by A. The resistance of the short is directly and the capacitance of the short is inversely proportional to distance S. The EQ. 7.9 shows the resultant impedance of such a short.

$$Z_{short} = \frac{R}{1 + j2\pi \times fRC} \approx R \qquad \text{(EQ 7.9)}$$

As can be concluded from EQ. 7.9, the impedance of the short is a function of the spacing S and also depends inversely on frequency and phase relationship of the two conductors. At low frequencies, the model of such defects is mainly resistive. However, above certain transition frequency (f_T) it becomes primari-

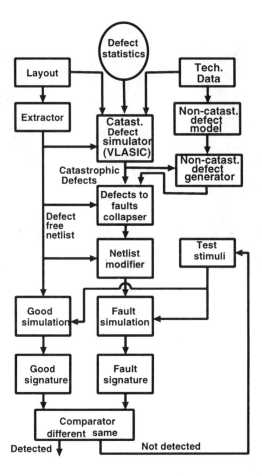

Fig. 7.4: A block diagram of the realistic defect based testability
methodology for analog circuits.

ly reactive. The transition frequency, f_T , depends on the defect geometry,
spacing S, and the resistivity and the permittivity of the insulating layer. A
particular soft defect may have very little impact on low frequencies but at
high frequency it may be significant. Fig. 7.3 shows a photograph of a non-
catastrophic short in metalization layer. The extra material defect reduces the
distance between two metal conductors giving rise to a high impedance bridging fault. However, for most applications and technologies the impedance of
the short can be approximated as purely resistive.

(b) Fault simulation environment: The block diagram of the environment is shown in Fig. 7.4. The process technology data, defect statistics and the layout of the circuit under investigation are simulation inputs. The defect statistics block contains the process defect density distributions. For example, probability of shorts in metalization is significantly higher than that for open defects in diffusion. A catastrophic defect simulator, like VLASIC [54] determines the realistic fault classes specific to the circuit and layout. VLASIC mimics the sprinkling of defects onto the layout in a manner similar to a mature, well-controlled production environment. The output of the simulator is a catastrophic defect-list. Analog circuits are also susceptible to non-catastrophic or parametric defects. Such defects are often called near-misses. We assume that such defects can occur at all places where catastrophic defects are reported by the defect simulator. However, pinhole defects are inherently parametric (high impedance) in nature. Therefore, only shorts and opens in various layers are considered for non-catastrophic defect generation. These defects are appended to the catastrophic defect list. The defect-list contains many defects that can be collapsed in unique fault classes. This process is carried out to find the likely fault classes in the layout. Subsequently, each fault class is introduced into a defect-free netlist for fault simulation. For the greatest accuracy, fault simulation is based upon a circuit simulator. The response of the fault simulator is called a fault signature. A fault is considered detected if the corresponding fault signature is different from the defect-free (good) signature by a predetermined threshold. If a faulty response does not differ from the good signature by the threshold, the fault is considered not detected by the stimulus and hence another stimulus is tried. This whole process is carried out for all faults.

A few things are worth highlighting in the above mentioned analog fault simulation methodology. First, unlike digital circuits, analog circuits lack the binary distinction of pass and fail. In fact, the decision of pass or fail in analog circuits is not clear cut. It depends on several variables including input stimulus, output measurement parameters (output voltage, I_{DD} current, etc.), circuit performance specifications and permitted environmental conditions (e.g., supply voltage, process, temperature, etc.). In other words, there is no absolute

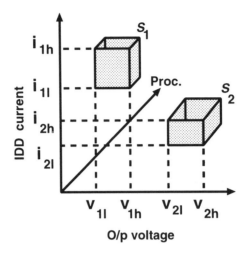

Fig. 7.5: The good signature spread.

reference for fault detection. A reference has to be evolved for a given circuit under given conditions. This generation of a reference is a tedious exercise and it should be created for each set of input stimuli. The impact of faults is measured against these set of references. Therefore, a reference response or good signature is a multi-dimensional space and the faulty circuit must exhibit a response outside this space to be recognized as faulty, at least by one of the test stimuli.

The graph in Fig. 7.5 illustrates this concept. In this graph, two axes form the primary output measurement parameters and the third axis forms an environmental condition (e.g., fabrication process spread). A set of graphs can be plotted essentially showing a possible good signature spread. The good signature spread (shaded area) is generated for each of the given test vector. A fault is considered detected by a given test vector if the faulty output response of the circuit lies outside the good signature space. For example in Fig. 7.6(a), the fault F_1 is detected by test vector S_1 with the output voltage measurement. However, it is not detected by the I_{DD} measurement since the faulty current lies within the good current spread. On the other hand, same fault is detected by test vector S_2 with output voltage as well as I_{DD} measurements. The infor-

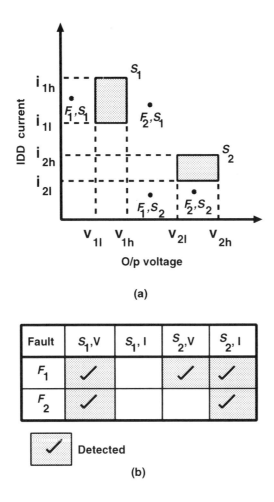

(a)

Fault	S_1, V	S_1, I	S_2, V	S_2, I
F_1	✓		✓	✓
F_2	✓			✓

✓ Detected

(b)

Fig. 7.6: (a) The fault detection; and (b) construction of a fault dictionary.

mation about fault detection is compiled into a fault dictionary D. Fig. 7.6(b) shows the fault dictionary. Rows of the fault dictionary show different fault classes (i.e., F_1 .. F_n) and columns show stimuli (i.e., S_1 .. S_t) with voltage (V) and current (I) as subfields.

Finally, for a structured analog DfT methodology to succeed, an effective and efficient test generation is of vital importance. Analog signals are time and amplitude continuous. Therefore, the concept of analog test vectors is not very well defined. For example, in digital domain a binary change in the input stimulus is termed as a change in test vector. These vectors are generated in a precise manner covering a predetermined fault set. However, in analog domain, often a test vector is defined as a set of input stimuli required for a particular measurement (specification). The parity between digital and analog test generation can only be restored if the basis for analog test generation is also a predetermined fault set. In this manner, true analog test vectors can be evolved. Furthermore, since all likely fault classes are known, in principle, simple test stimuli can detect the presence (or absence) of a defect.

7.5 A Case Study

We use a class AB stereo amplifier as a vehicle to examine the effectiveness of this methodology. This chip is mass produced for consumer electronics applications. Owing to high volumes and low selling cost, it is desirable to cut down the chip test costs and at the same time maintain quality of the shipped product. It is a three stage amplifier. The first and second stages are completely differential in nature. The outputs of the second stage feed the output stage which drives a load of 32 ohms. It was designed in a standard 1.0 micron single poly double metal CMOS process. The chip contains two identical amplifiers (channels A and B) and a common biasing circuit. Since both the channels of the class AB stereo amplifier are identical, only one amplifier is considered for testability analysis.

7.5.1 Fault matrix generation

VLASIC was utilized to introduce 10,000 defects into the layout. Since most of the defects are too small to cause catastrophic defects, only 493 catastrophic defects were found. These defects were further collapsed into 60 unique fault classes. Table 7.1 shows the relevant information about various fault classes due to catastrophic defects. A catastrophic short in metal layers was modeled as a resistor with nominal value of 0.2 ohm. Similarly shorts in poly and diffusion layers were modeled with a resistor of 20 and 60 ohms, respectively. Extra contact and via were modeled with a resistor of 2 ohms. Thick

Table 7.1: Catastrophic fault classes and their fault models.

Defect	Number	%	Model (Ohm)
Shorts	22	37	0.2,20,60
Extra contact	10	17	2
Oxide pinhole	15	25	2k
Gox. pinhole	7	11	2k
Junc. pinhole	6	10	2k
Open	0	0	--
Total	60	100	

oxide defects were modeled as a resistor of 2k ohms. The gate poly is doped n-type and all gate oxide shorts occurred in n-channel transistors causing shorts between the gate and the source or drain of transistors. Therefore, such shorts were non-rectifying in nature and hence were modeled as a 2k resistor. The n-channel transistor is more susceptible to gate oxide shorts and most of the gate oxide shorts are likely to occur between the gate and source or drain [43].

As mentioned before, soft faults were evolved from the hard fault data. Soft faults were generated at locations of shorts and opens in interconnect, contacts and vias. Therefore, 32 soft fault classes (first 2 rows of Table 7.1) were evolved. Rodriguez-Montanes [32] reported that the majority of bridging defects are below 500 ohms. Therefore, the resistance of non-catastrophic defects was chosen as 500 ohms. The capacitance was calculated from the technology data keeping the spacing between the defect and the conductor (s) as 0.1 micron. The computed value is 0.001 pF.

All catastrophic defects in the defect-list generated by VLASIC were shorts in nature caused by extra material, oxide pin-holes or extra contacts. None of the defects caused an open circuit. However, given the defect densities for various defects in the fabrication process, it was hardly surprising. The shorts in the back-end of the process constitute the majority of the spot defects. Furthermore, the layout geometries in analog circuits are often non-minimum size and multiple contacts and via contacts are utilized to reduce the contact resistance. All this put together made occurrence of an open in the given layout less probable. However in real life, the nature of above mentioned defects can vary a great deal and hence no simulation can claim to be exhaustive. Nevertheless, these numbers are consistent with the resistivity of respective layers and the published data. Furthermore, for such an analysis, the order of defect resistance is more important than the absolute value.

7.5.2 Stimuli matrix

For this case study, we divided test signals in three categories: (i) DC stimuli, (ii) 1 kHz sinusoid stimuli and (iii) AC stimuli. Often the analog circuit function depends on the continuous operation of all sub-blocks. Therefore, it is quite likely that a catastrophic fault would change the DC operating point of the circuit and hence will be detected by a DC test stimuli. This may also hold true for some high impedance non-catastrophic faults as well. A lower frequency sinusoid was chosen since many fault classes may not be excited under DC conditions. Finally, AC stimuli were chosen because the impact of many non-catastrophic faults is frequency dependent and it is worthwhile to analyze the frequency response as well.

For the simulation of fault classes, the amplifier is put into the configuration shown in Fig. 7.7. A load of 32 ohms is placed on the output with a DC blocking capacitor. A 2k ohms feedback resistor is placed between the output and the negative input. Furthermore, a full load is put at the output. Before proceeding to fault simulation, the defect-free response is compiled. A fault is considered detected if defect-free and faulty responses differ by at least 1 volt for output voltage measurement or by 0.5 mA for supply current measurement. For AC analysis a fault is considered detected if it modifies the frequency response by 3 dB. For the DC analysis the positive input is held at 2.5 volts

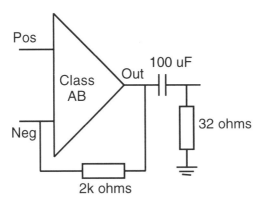

Fig. 7.7: Fault simulation configuration for the Class AB amplifier.

and the negative input is swept from 0 to 5 volts. The output voltage and current drawn from VDD are sampled when negative input is 1,2, and 3 volts, respectively. Similarly, for low frequency transient analysis, a 1 kHz sinusoidal signal is applied and root mean square (rms) values of output voltage and I_{DD} are calculated. For AC analysis different frequency signals on the negative input are applied while the positive input is held at 2.5 volts. In this configuration, the gain of the amplifier is measured.

7.6 Results

7.6.1 Simulation results

Fig. 7.8(a) shows the result of fault simulation for catastrophic faults. In this figure results are independently segregated according to the mode of the analysis. On the X-axis, the mode of analysis means the type of input excitation. The detection mechanisms are represented by output voltage, supply current and gain of the amplifier. The Y-axis shows the percentage of faults detected by each mode of analysis and detection mechanisms independently. For example, DC voltage detection of a fault means that the particular fault was detected by output voltage measurement when input excitation was DC. The third

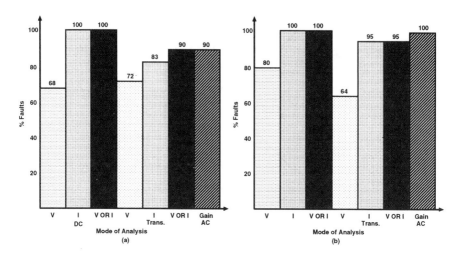

Fig. 7.8: Fault simulation result for (a) catastrophic defects and
(b) non-catastrophic defects.

column in DC analysis shows how many faults were detected either by DC
voltage or by DC current. Therefore, it is the logical OR of first two columns
of the same analysis. Similarly, the third column in transient analysis is the
logical OR of first two columns of transient analysis.

In DC analysis, 68% of the faults were detected by output voltage. However,
the current is a better measure for fault detection and all faults were detected
by it. Needless to say that when both detection mechanisms, voltage and current, are considered together, all faults were detected. Though the voltage detection of faults in transient analysis is higher than that of DC analysis, the
results, in general, were less encouraging. The 72% faults modified the voltage signature of the device and 83% faults modified the current drawn from
VDD. When both mechanisms were considered together, 90% of the faults are
detected. Lower than expected performance of transient analysis compared to
DC analysis can be attributed to the fact that transient defect-free current or
voltage signature is sinusoidal and comparison of two sinusoids on tester (or
in simulation) is more difficult than the comparison of two DC currents.
Therefore, in transient analysis, fault detection is carried out manually. If a

fault modified the response more than the determined threshold, it is considered detected. In AC analysis, 90% of the faults modified the frequency response of the circuit and hence are detected.

Fig. 7.8(b) shows the fault simulation results for non-catastrophic faults. The effectiveness of current in DC analysis for fault detection is once again demonstrated. However, for such defects, gain of the amplifier is also an important detection mechanism. All faults were detected by both analyses. Given the model of these non-catastrophic faults (500 ohms) it was expected.

7.6.2 Silicon results

Conventionally, devices are tested by verifying a set of DC and AC specifications. The DC specifications include input offset voltage, input bias current, common mode voltage range, output voltage swing, output impedance, output current, etc. The AC specifications include total harmonic distortion (THD), signal to noise ratio (SNR), slew rate, output power, etc.

A set of 18 passed devices and 497 failed devices with the conventional test process are selected. The passed devices are selected to observe the spread of good signature compared to simulated thresholds. The comparison is shown in Table 7.2. The values shown outside brackets represent actual (measured) and inside brackets represent simulated thresholds for pass/fail. The actual voltage spread is much smaller than the simulated threshold, however, actual current spreads are at least an order of magnitude larger than the simulated threshold. One of the explanations for high current spread is that for this experiment the device is excited in a different manner compared to its normal usage. Therefore, current spread in this configuration was not controlled.

The performance of the defect-oriented tests on failed devices is shown in Fig. 7.9. For channel A, DC and transient voltage as well as the gain measurements caught all faulty devices. The performance of current measurement was less satisfactory. This difference between simulated and silicon results was due to high current spread in defect-free silicon signature. The current measurement

Table 7.2: Good signature spread for manufactured chips.

	DC	Trans.	AC
Voltage	0.15 (1) V	0.1 (1) V	2 (3) dB
Current	9 (0.5) mA	5 (0.5) mA	—

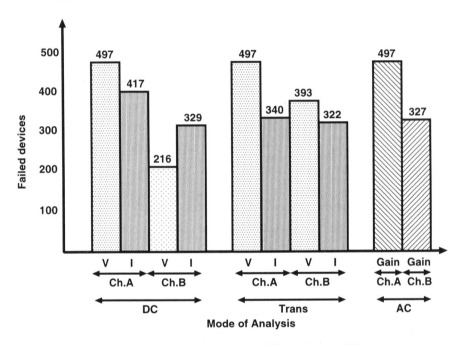

Fig. 7.9 : First silicon result for Class AB amplifier.

with DC was more effective than that with transients. On the other hand, for channel B, DC voltage caught fewer defects compared to DC current measurements. However, transient voltage was more effective than the transient current measurement. The gain measurement of channel B detected 327 faulty devices. In general, channel A failed more often. No failure analysis was carried out to determine the causes of the difference between channels. Probably subtle layout differences between channels are the reason for different failure rates.

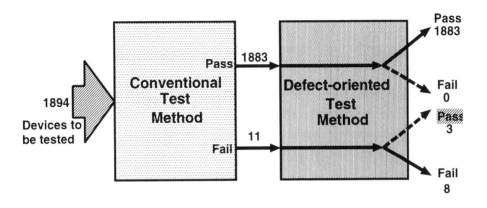

	measured offset	test limit	measuredS/N ratio	test limit
# 1	91.4 mV	<90.0 mV	100.2 dB	>101 dB
# 2	94.5 mV	--	100.1 dB	--
# 3	96.4 mV	--	99.7 dB	--

Fig. 7.10 : The results of the second experiment over the Class AB amplifier.

Subsequently, the defect-oriented test method was put into the production test environment along with the conventional test method. A test program was evolved in which devices were first tested with the conventional test method and then with the defect-oriented test method. This exercise was carried out to determine the effectiveness of the defect-oriented method with respect to the conventional test. A total of 1894 devices was tested. The yield of the device was very high and only 11 devices failed the conventional test. Out of these 11 devices 3 did not fail the defect-oriented test. These three devices were again tested with the conventional test method. Table in Fig. 7.10 illustrates causes of failures of these devices by the conventional test method. Their failures were marginal and very close to the specification limits. The likely origin of such failures lies in inherent process variation and not in spot defects. A test methodology with spot defects as the basis can not ensure detection of such faults. Improved control of process is one possible solution to reduce such escapes.

The difference in simulation and actual results is quite apparent. Several factors contribute to differences: (i) A finite sample size of defects does not cover whole spectrum of possible defects. (ii) A defect may have many possible impedances. A simulation for all observed defects and impedances is beyond the capabilities of any state-of-the-art simulator. (iii) A circuit simulator has limited capability in simulating actual silicon behavior. (iv) As mentioned before, higher silicon current spread limits the fault detection capability of current measurement method. Therefore, DC as well as transient voltage measurements appear to be more effective for fault detection in devices. (v) Finally, the defect-oriented test methodology is based upon spot defects and global or systematic defects and process variations are not taken into account. Such non-modeled faults are also possible due to differences between simulation and the actual test results.

From the second test experiment [34,35] two broad conclusions were drawn: (i) simple tests can detect catastrophic failures, however, detection of some subtle failures is uncertain, and (ii) the number of failed devices is not sufficient to draw any meaningful conclusion about the method's applicability in catching real life faults. More test data, especially on faulty devices, is needed to substantiate claims of IFA based tests. We report a relatively large experiment over the same Class AB amplifier devices with the objective to find the effectiveness of the test method on catching real life failures [36].

Fig. 7.11 illustrates this experiment. A total 3270 rejected samples of Class AB amplifier were gathered from a total of 106,784 devices tested by the conventional test method. Only failed devices (3270) were considered for further testing. These devices were tested with the IFA based test method. Out of this lot, 433 devices passed the test. These passed devices from the IFA based test method were once again tested with conventional test method. Results of this test were following: (i) 51 devices passed, the test, and (ii) rest of the devices (433-51=382, or 0.4% of total tested devices) failed the test again. These failed devices (382) were subjected to a detailed analysis.

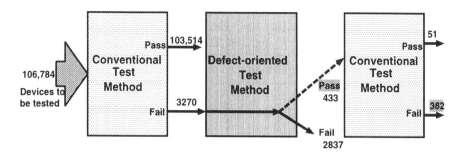

Fig. 7.11: The results of the third experiment on Class AB amplifier.

7.6.3 Observations and Analysis

Table 7.3 shows the result of the analysis of 382 failed devices. The input off-set voltage specification contributes to the largest number of failures (182 or 47.6%) that could not be caught by the IFA based test method. The total harmonic distortion (THD) specification contributed to the second largest segment of undetected failures (123 or 32.2%). Similarly, SNR measurement failed 20 devices (5.2%). These three categories of failures contribute to the bulk (85%) of the failures that could not be detected by the IFA based test method. These failures can be attributed to un-modeled faults in the IFA process. For example, a differential amplifier has an inherent offset voltage that is the source of non-linearity in its operation. Often this offset voltage is minimized by transistor matching, layout, trimming and compensation techniques. Besides process defects, several other factors can increase the offset voltage. The increased offset voltage (within specification limits) increases non-linearity, reducing the SNR ratio of the amplifier. In the table of Fig. 7.10 the device with the highest input offset voltage shows the lowest SNR and the device with the lowest input offset voltage shows the highest SNR.

In general, tighter the parametric specifications of an analog circuit, less effective IFA based test is likely to be. This is because natural process variations with higher parametric requirements will contribute to larger number of de-

Table 7.3 : Analysis of devices failed in third silicon experiment.

Number	Percentage	Failure mechanism
2	0.5	Open/Short
41	10.7	Supply Current
182	47.6	Offset Voltage
1	0.3	Output Voltage swing DC
10	2.6	Common Voltage
0	0.0	Output Voltage AC
20	5.2	S/N ratio
123	32.2	THD at 1 kHz
0	0	X-talk
3	0.8	Ripple rejection
382	**100**	

vice failures. Since these are un-modeled faults, the effectiveness of IFA based test is lowered. Furthermore, the effectiveness of a process defect based yield loss model diminishes significantly with increasing parametric requirements. Therefore, a test based solely on process defects is not sufficient for ensuring the specifications of the device with high parametric specifications.

7.6.4 IFA: Strengths and Weaknesses

On the basis of the above experiments, we make the following comments regarding strengths and weaknesses of IFA for analog circuits. Some of these comments are specific to the Class AB amplifier and others have general applicability.

- An IFA based test method is based upon process defects. This is in contrast with the conventional, specification based, analog test method. The IFA based test method is structured and, therefore, has a potential for quicker test generation. Though IFA based test generation requires considerable effort and resources, it is faster than the specification based test generation.

- The IFA based tests are simpler and their requirements for test-infrastructure are substantially lower compared to the specification based tests. Therefore, majority of such tests can be carried out with inexpensive testers. A vast majority of faults is detected by simple, DC, Transient and AC measurements. For example, the Class AB amplifier devices are tested with a combination of IFA based test and limited functional test. The combined test method results in an estimated saving of 30%.

- The number of escapes (382) of the IFA based method amounts to 0.358% of tested devices (or 3,580 PPM). Clearly, it is unacceptably high. A limited specification test, as mentioned above, with IFA based test may be advantageous in quality improvement while test economy is maintained.

- The number of escapes can be reduced by a rigorous control of the fabrication process. The basis of IFA is a given set of process defects. However, this basis is not absolutely fixed because of the process dynamism. A new defect type may be introduced into the set if the process is unstable or improperly monitored. A better process control (higher Cp and Cpk) will increase the effectiveness of the IFA based test.

- Effective test generation and limit setting is of crucial importance to the success of IFA based testing. For example, when supply current measurements were implemented in IFA tests, a substantial amount of devices (41) passed the test but failed the supply current test in the conventional test method. This is because the test limits in IFA based test are determined more or less arbitrarily. The measured current on 187 good samples suggests that test limits should be more stringent. The same holds true for other detection-thresholds. Setting of 1 V or 3 dB thresholds for fault detection is not stringent enough to ensure high parametric fault requirements for the amplifier. More research is needed for test pattern generation and threshold settings.

- Design insensitivity to process variations also contributes toward the effectiveness of the IFA based test vectors. IFA based tests are ill-suited for design characterization.

- The Class AB amplifier is an audio amplifier with very tight parametric specification and relatively small number of transistors. IFA based test methods are more successful for circuits or ICs where the parametric specification are relatively relaxed and functional complexity is high. For such complex ICs, functional testing is not enough and IFA test may form the main segment of testing. On the other hand, for high performance analog ICs the IFA based simple test may form the basis of wafer sort, rejecting all potentially defective devices. The subsequent limited functional test will be applied only to potentially good devices. The combination of these two will not only improve the economics of testing but will also result in better quality of tested devices.

- Quantifying fault coverage of a given set of test vectors for an analog circuit is an unexplored area. The IFA based test generation provides a methodology by which test vectors and design can be fault graded. Once, fault coverages of different tests are known, ordering of tests may improve test economics. Tests that do not contribute to fault detection may be discarded. Furthermore, the impact of test vectors on outgoing quality can also be quantified.

- IFA based test method is limited by the availability of CAD software tools and requires high computer resources in terms of CPU power and data storage. A substantial analysis effort is needed before an IFA based test method can be implemented. Furthermore, due to computational and CAD tool related constraints only cells and macros can be analyzed. Therefore, ideally this analysis should be carried out in the design environment on a cell by cell basis. A bigger design should be partitioned into suitable smaller segments for analysis.

7.7 IFA based Fault Grading and DfT for Analog Circuits

One of the major issues faced in analog testing is how to quantify the existing test methods (e.g., functional tests) against the manufacturing process defects. In analog circuits, as we see from experiments, the functional or specification based testing cannot be eliminated completely in favor of simple DC or AC tests for circuits with tight parametric specifications. Furthermore, popularity of mixed-signal devices has compounded the analog test issues. The testing of analog blocks in a sea of digital logic is becoming an increasingly difficult task. Two major issues pose difficulties. Firstly, limited controllability and observability conditions for analog blocks increase test complexity and cost. Secondly, in digital domain, a large number of test methods (e.g., functional, structural, I_{DDQ}) and DfT techniques (e.g., scan path, macro test) are available for quantifying and improving the fault coverage. Furthermore, automatic test pattern generation (ATPG) techniques have reduced the test generation cost for digital circuits significantly. Analog testing lacks such tools and techniques. Therefore, analog testing is becoming a bottleneck in testing of mixed-signal ICs in terms of cost and quality.

The quality of the test, and hence that of the tested device, depends heavily on the defect (fault) coverage of the test vectors. Therefore, it is of vital importance to quantify the fault coverage. Since the fault coverage of the test vectors on various building blocks of a design is largely unknown, the benefits of any DfT scheme can not be ascertained with confidence. Furthermore, one can not conclude where DfT is needed most. This lack of information has resulted in the abuse of digital DfT schemes in the analog domain and is probably one of the important contributing factors in the demise of analog DfT schemes. We demonstrate how the IFA technique can be exploited to fault grade the given (conventional) test vectors. Once, the relative fault coverage of different blocks is known by given test vectors, an appropriate DfT scheme can be applied to the areas where fault coverage of existing test methods is relatively poor. This is demonstrated with an example of a flash A/D converter.

7.7.1 A/D Converter Testing

An A/D converter is normally tested for DC and AC performance. The DC tests typically test for offset voltage and full scale errors. Static differential non-linearity (DNL) and integral non-linearity (INL) measurements are performed by slowly varying the input signal such that the DC operating point is reached for each measurement. On the other hand, dynamic tests are performed to test for dynamic range, conversion speed, SNR, dynamic DNL, dynamic INL, bit error rate (BER), etc. These dynamic specifications are often tested by performing BER measurement, code density measurement (CDM), beat frequency measurement or SNR or THD measurement.

The code density measurement is an effective way of testing A/D converters. The static DNL and INL of the converter can be computed from this measurement. At the input of an A/D converter a waveform is applied. The amplitude of this waveform is slightly greater than the full scale value of the converter. As the waveform traverses from zero to full amplitude, different output codes appear at the output of the A/D converter. For an accurate measurement at least 8 to 16 codes per level are needed [18,22]. This is achieved by repeating the test for a number of cycles of the waveform. Often a triangular waveform is applied because then every code should have equal density. If a larger or a smaller number of codes is found in the CDM, it shows the presence of poor DNL. A fault is considered detected by CDM if it resulted in more or fewer (pre-specified criterion) occurrences of a given code.

For the SNR, THD and SINAD measurements, a sine wave is applied at the input of a converter and output codes are measured. In order to randomly distribute the quantization error over the measurement, the ratio of signal frequency to the sampling frequency is given by EQ. 7.10.

$$\frac{F_{signal}}{F_{sample}} = \frac{M}{N}$$
(EQ 7.10)

where M and N are mutually prime integers and N is the number of samples taken. Mahoney [17] calls it M/N sampling.

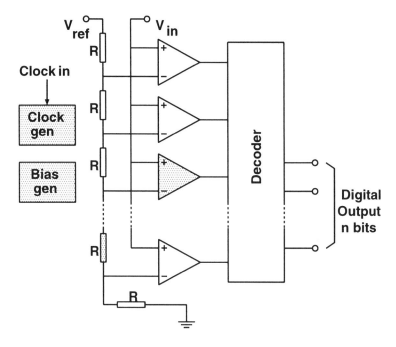

Fig. 7.12 : Block diagram of a flash A/D converter.

7.7.2 Description of the Experiment

An 8-bit flash A/D converter [28] is utilized for this experiment. However, the 8-bit flash A/D converter is too complex even for a fault-free simulation at the Spice level. Nevertheless, for the accuracy of the analysis, the circuit level simulation is considered to be an absolute requirement. Therefore, a 3-bit model of the converter at the Spice level was made. This model has only 8 (instead of 256) comparators which could be simulated in reasonable time.

IFA is performed over all basic cells of the converter to find out the fault classes. The 3-bit model has 8 comparators, one clock generator, one biasing generator and one reference ladder. For the fault simulation purposes, a fault is introduced in one of the instances of a basic cell in the model. This is shown in Fig. 7.12. For example, faults were introduced one by one in only one out of eight comparators at a time. This was under the single fault assumption in the design. The complete A/D model is simulated for all comparator faults. Once, all faults for the comparator were simulated, faults were introduced in another cell (e.g., clock generator or biasing circuit). In this way, all likely faults in the circuit were simulated.

7.7.3 Fault Simulation Issues

In analog circuits, fault simulation is a laborious exercise. For each fault, a separate simulation is run. Relatively high degree of human interaction is required for analog fault simulation. Furthermore, there are analog fault simulation issues that should be carefully addressed.

The simulation environment is considerably slower from that of a tester. A test that takes fraction of a milli-second on a tester to perform may cost several minutes in simulation environment at Spice level. Furthermore, since fault simulation is to be performed over the complete fault set, the total time for the analysis may become prohibitively large. Due to time constraints, for the CDM, we reduce the average number of codes to 5 and applied a slow ramp so that every 5th clock cycle a new output code is generated. In other words, for 5 clock cycles, the fault-free converter is supposed to have the same output code. Even then, single fault simulation over 3-bit A/D model took 8 CPU (HP 700) minutes with CDM test. We considered a fault detected by CDM if it resulted in more than 7 or less than 3 occurrences of a given code. The SNR, THD and SINAD measurement takes 45-50 CPU minutes for single fault simulation. We selected SINAD instead of SNR as fault detection criterion. SINAD is defined as the signal to noise plus distortion ratio. The fault simulation using BER could not be performed since it would take even more time than the SINAD test. The DNL and INL measurement were carried out using the data of SINAD tests.

Secondly, we utilized a production test system, MixTest [23], to compute DNL, INL and SINAD of fault simulations. A fault is detected by DNL, INL, or SINAD if the computed value differs from the golden device simulation by a predetermined threshold. The sampling frequency of the converter was 20 MHz. To randomly distribute the quantization errors, a fraction of the sampling frequency $\{(31/128)\times20$ MHz$\}$ was selected as the input frequency.

Table 7.4: The fault simulation results on the flash A/D converter.

Test Cell	Code density measurement	SINAD	DNL	INL	Undetected faults
Comparator	112 (157)	19 (45)	25 (45)	21 (45)	17
Clock generator	32 (59)	10 (27)	14 (27)	10 (27)	11
Bias generator	16 (50)	8 (34)	14 (34)	9 (34)	18
Ref. ladder	16 (19)	1 (3)	1 (3)	1 (3)	2

Thirdly, the setting up of the thresholds for fault detection in simulation environment must be done carefully. For example, the criterion of 1 LSB for DNL and INL measurements is no longer valid for a 3-bit model of an 8-bit converter. For the original converter, 1 LSB is 2V/256=7.8 mV. For a 3-bit model, 1 LSB amounts to 2V/8= 250 mV that is substantially greater than 7.8 mV. Owing to the constraints of simulation environment, we selected 0.1 LSB or 25 mV as the detection threshold. Though, this may be a little conservative, we assumed that if a fault is detected against a relaxed threshold criterion, it will certainly be detected in the production environment against tighter limits. Similarly, for a 3-bit converter, theoretical SNR should be 19.82 dB and the theoretical SINAD should be somewhat lower than this value. The SINAD for the 3-bit model in fault-free simulation was found to be 18.05 dB. Once again, we took conservative values for the fault detection. A fault is considered detected by SINAD measurement if the SINAD of the converter was less than 17.5 dB.

7.7.4 Fault Simulation Results

Table 7.4 compiles results of the fault simulation of 3-bit A/D converter model. In the comparator, 157 fault classes were simulated with the CDM test. A set of 112 faults were detected by that test. Owing to large fault simulation time, only those faults that are not detected by CDM (157-112=45) are simulated for SINAD, DNL and INL tests. This is further justified by the fact that CDM test is a simplified version of SINAD, DNL and INL tests. Therefore, if a fault is detected by CDM, it is likely to be detected by these tests. The DNL test was most effective (25/45) in catching rest of the undetected faults in the comparator. INL and SINAD, detected 21/45 and 19/45 faults, respectively, in the comparator. Nearly, 11% of the faults (17/157) in the comparator were not detected by any of these measurements. In the case of the clock generator, 59 fault classes were simulated. The CDM could detect 32 of the simulated faults. The performances of DNL, INL and SINAD, for the remaining undetected faults (27) were relatively poor compared to the comparator. As a result 11 of the 59 clock generator faults remained undetected. The performance of conventional tests was the poorest on the bias generator. A total of 50 fault classes were simulated in the bias generator. The CDM could detect only 16. The performance of DNL was marginally better. It detected 14 out of the remaining 34 undetected faults. On the whole, 36% of the total faults remained undetected. In the reference ladder, 19 fault classes were simulated. The CDM was an effective test and detected 16 fault classes. DNL, INL, and SINAD detected 1 fault class out of the 3 undetected fault classes and 2 fault classes remained undetected.

7.7.5 Analysis

Nearly 20% of the faults in the clock generator and 36% of the faults in the bias generator are not detected by the commonly used A/D specification tests. On the other hand, nearly 90% of the faults in the comparator and resistor ladder network are detected by these tests. The difference in fault coverage is easy to explain. Most of the conventional specification (conversion speed, SNR, DNL, INL, BER, CDM, etc.) tests are targeted toward faithfulness of the data path (i.e., analog input, reference ladder, comparator, decoder, digital output). There is hardly any test that explicitly covers the control path (i.e., clock and the bias generators). These blocks are assumed to be tested implicit-

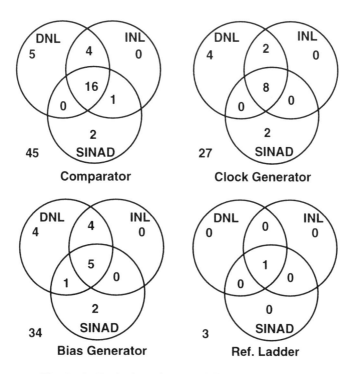

Fig. 7.13: Fault detection capabilities of different tests.

ly. Poor controllability and observability are other reasons for undetected faults in these blocks. The outputs of these cells are not directly observable. If the faulty bias or clock generator output causes the comparator to behave in an incorrect manner, then the fault is detected by the tests. However, the faults that modify the behavior of the control path only marginally are hard to detect and require testing of the complete dynamic ranges of input amplitude and frequency. Comparators are often designed to withstand the parametric variations of the clock and biasing to optimize the yield. Such a design has a fault masking effect.

Different specification tests differ in fault coverage. The relative fault coverages of different tests are illustrated in Fig. 7.13. As explained in previous subsection, we only considered faults not detected by the CDM test. Though most faults are detected by all tests (DNL, INL, SINAD), DNL is the most effective test. Some faults are only detected by SINAD. However, INL does not detect any fault not detected by the other two tests. Nevertheless, we should keep in mind that INL is not a redundant test. INL is an effective test for detecting parametric variations in the reference ladder. For example, it covers those parametric variations that do not cause appreciable shift in the DNL but affect the whole reference ladder.

7.7.6 DfT Measures

An important question is how the fault coverage can be improved without sacrificing the performance of the converter. Measurement of the quiescent current (I_{DDQ}) may be one solution. Unlike digital circuits, the I_{DDQ} of an analog circuit is not in the sub-μA range. Therefore, its detection capability is limited. Alternatively, The A/D converter should be designed such that all high current dissipating paths are either switched off or bypassed for I_{DDQ} testing. Thus, I_{DDQ} test can be made an effective test method in fault detection. However, the design of such a converter requires a non-trivial amount of effort.

There are innovative voltage DfT techniques that do not cause performance degradation and improve fault coverage. For example, a DfT scheme for the clock driver is explained in Fig. 7.14 and Fig. 7.15. Fig. 7.14 shows 4 clocks generated from the clock generator. These signals are digital in nature but their timing relationship with each other is extremely important for a correct function of the A/D converter. For a DfT solution, we exploit the knowledge of pre-defined timing relationship between different clocks. Typically, a large number of faults degrades the timings of clock signals, if we take a logical AND of these signals (Fig. 7.15), we get an output pulse whose position and width are known. Most faults causing timing and/or stuck-at behavior will be detected by the measurement of pulse position and/or width. The number of pulses within the clock cycle and their position from CLK_IN can be the fault detection criteria. Then, more than 95% of the faults influenced output(s) of the clock generator and were detected within 2 clock cycles. This test method detects faults quickly and provides the diagnostic information. There can be a variety of implementations to extract different attributes of periodic signals. It costs approximately 10 logic gates for the particular implementation. The number of gates is a trade-off between the required diagnostics and the cost of

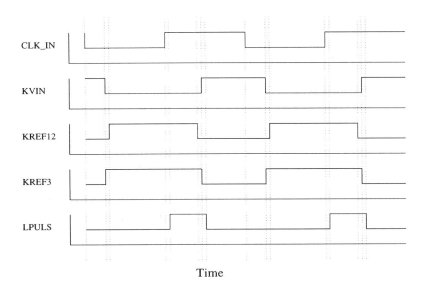

CLK_IN

KVIN

KREF12

KREF3

LPULS

Time

Fig. 7.14: Input clock signal and various generated clocks in the analyzed flash A/D converter.

implementation. The number of gates can be reduced if the critical signal spacing requirements are known in advance. For example, signals KVIN and KREF12 should be non-overlapping. Hence, only the critical timings are generated by the Boolean operations.

Similarly, a DfT solution for faults in the bias generator is shown in Fig. 7.16. Bias generator provides a set of stable bias signals to the comparator. In the case of the flash A/D converter 4 bias signals are generated. These are stp1 (3.2V), stp2 (3.2V), stp3 (3.4V), and Vbias (2.3V). Each biasing voltage is applied to a p-channel transistor that is individually gated by an n-channel transistor. This scheme allows measurement of quiescent current through individual or multiple paths. Defect-free quiescent currents through the components can be computed. If the presence of a defect in the biasing network influences the voltage level of any of the biasing signal, it is translated into a current that can be measured. Nearly 80% of the bias generator faults are detected by these simple measurements. It is worth mentioning that the popular

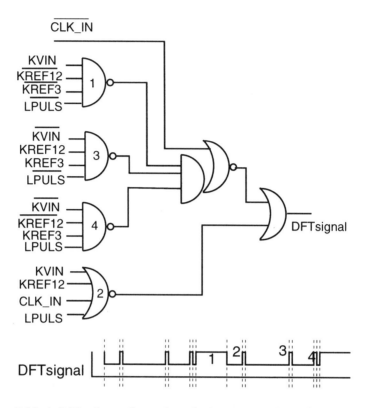

Fig. 7.15: A DfT scheme for testing clock generator faults in the analyzed flash A/D converter.

and expensive (conventional) test method detects only 60% of these faults.

Alternatively, the available infra-structure of the same A/D converter (comparators, etc.) may be utilized to determine the voltage of various internal biasing and clock signals. The same idea can be extended to test external analog blocks preceding the A/D converter. Typically, these blocks are noise shapers, filters, amplifier, etc. The basic idea of this DfT concept is illustrated in Fig. 7.17. In the test mode, instead of connecting the normal signal (V_{in}) to comparators, the desired internal signals of the converter are selectively applied to the comparators. The reference voltage (V_{ref}) is appropriately applied such that the comparators can compare the applied signals with optimum accuracy. These measured signals may include time invariant and/or analog signals (i.e.,

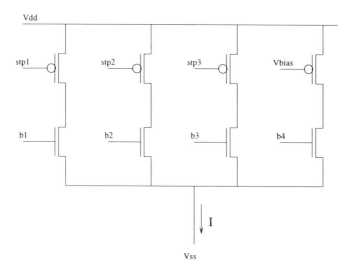

Fig. 7.16: A DfT scheme for testing bias generator faults in the analyzed flash A/D converter.

biasing signals, internal, external nodes of the A/D converter) or digital (Boolean) signals (clocks). The test mode signal is converted into the digital output that can be interpreted against pre-determined values. The test method may also be used for design verification and diagnostic purposes. Furthermore, the method may used be for in-system testing or BIST applications.

Fig. 7.18 illustrates a schematic of the test controller. It contains a decoder that, depending on the input code (Test), connects a given signal to the comparators. In the figure, signals T1, T2, and T3 form the decoded address. In embedded applications, these signals may be applied through a scan chain. Alternatively, at the cost of adding a simple counter, they may be generated within the converter. Let P1, P2, ... be the periodic signals and let B1, B2, ... be the biasing or time invariant signals for the A/D converter. For testing of an arbitrary analog biasing signal, say B1, the appropriate address value for the test decoder is applied such that B1 is connected to the bank of comparators. Biasing signals are transmitted through transmission gate pairs to minimize signal degradation. Furthermore, biasing signals are DC levels and the effect of transmission gate resistance (R_{on}) in test mode should not be of major sig-

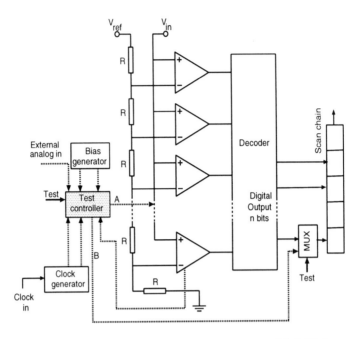

Fig. 7.17: A full flash A/D converter block diagram with DfT features.

nificance. Alternatively, signal degradation can be characterized with a signal of known amplitude. The output of the converter gives a digitized value of the bias signal. As mentioned previously, biasing signals are often not tested explicitly and biasing stability with respect to the temperature and other environmental conditions is rarely examined. A biasing circuitry, under the extreme conditions may acquire an oscillating behavior that is very difficult to test. This DfT scheme allows to test the conditions quickly and unambiguously. In a similar fashion other internal analog signals of the A/D converter may be tested. The same method my be utilized to test external analog blocks or analog signals inside them individually. However, when testing a high speed analog signal care should be taken to optimize the test for least signal degradation in the test mode.

Testing of digital signals does not require an A/D converter. However, for good functioning of an A/D converter, precise clocking is vital. A way of testing internal clocks is shown in Fig. 7.17 and Fig. 7.18. Let us assume that clock P1 is to be tested. An appropriate address value for the test decoder, T, is

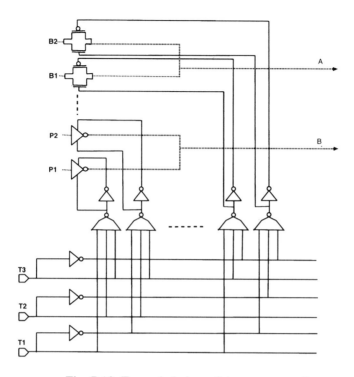

Fig. 7.18: Expanded view of the test controller.

applied. The same test address also enables the output multiplexer such that a part of the digital output contains the clock information that can be shifted out via scan chains. Alternatively, for digital signals, the multiplexer and part of the test controller may be incorporated into the output decoder of the A/D converter such that in the test mode the digital outputs shows various critical digital signals of the comparator. The implementation is simple and does not require further explanation.

7.8 High Level Analog Fault Models

Defect oriented fault analysis of analog building blocks is a tedious task due to lack of tools and high computation time. Spice level fault simulation often requires prohibitively large simulation time. Therefore, there have been attempts to create higher level fault models based upon realistic defects such that the

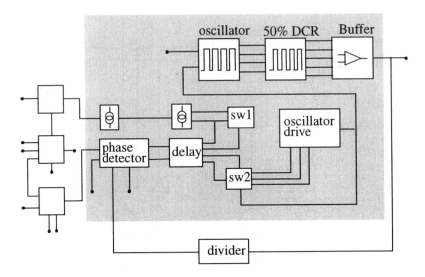

Fig. 7.19: Partitioning of the PLL into macros.

simulation time may be reduced [1,10,16]. Harvey et al. [10] carried out a defect oriented analysis of a phase locked loop (PLL). Fault free simulation of
PLLs takes enormous CPU time. Therefore, higher level fault models were
utilized. The PLL was also divided into several macros (Figure 7.19) for
which simulation models at behavior-level were developed. Although the simulation time per macro was reduced significantly as can be seen in Table 7.5, it
still took several hours of CPU time to simulate the locking behavior. Circuit-
level simulation was not feasible.

For the IFA analysis, faults were inserted into each macro in turn. To ensure
correct fault behavior, the macro being analyzed was replaced by its circuit-
level description. Most faults caused a hard failure that was already identified
in the simulation of the macro being analyzed. Only a few of the faults had to
be simulated in a functional way, including all models of the other macros. A
complete analysis revealed that the functional test, comprising locking time
and capture range measurements, will detect about 93% of the faults. Alternative tests were evaluated with respect to customer quality requirements. The
remaining 7% faults can be detected by using power supply voltage levels out-

Table 7.5: Comparison of circuit level and behavioral level simulation times.

Block	Circuit CPU time (s)	Model CPU time (s)	Speed advantage factor
Oscillator	3082	134	23
50% DCR	803	527	1.5
Buffer	96	4	24
Phase detector	373	14	27
Delay	82	5	16
Current mirror 1	76	3	25
Current mirror 2	106	3	35
Current switch 1	42	3	14
Current switch 2	72	2	36
Oscillator drive	58	6	10

side the specified operational range, that extends from 4.5 to 5.5 Volts. Application of a supply voltage of 3 Volts changes circuit sensitivities [6] and thus enables the detection of other faults [10]. Such tests should only be used when the fault-free response can unambiguously be identified.

In a similar manner, an anti side tone module of the PACT IC, used for telephone applications (PCA1070 [2]) was subjected to fault analysis. This module consists of two programmable switched capacitor filters in a feedback loop (see Fig. 7.20). Being programmable, the IC can be used in various countries with different statutory requirements, but the necessity of the repetition of functional tests for different settings increases the time required for testing.

Goal of the analysis of the anti side tone module was to evaluate the effectiveness of the functional tests used in production testing. Again, a macro oriented divide and conquer approach was used. For the analysis of the ZS-filter, other macros were modeled at a level optimized in terms of simulation speed. The

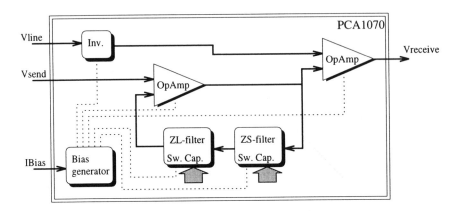

Figure 7.20: The anti side tone module of the PCA 1070.

assembly of all macros together was used in the fault simulation.

The most complex macro is the ZS-filter, a programmable switched capacitor filter, whose programmed digital setting determines the number of switched capacitors that is activated. It takes a long time to simulate this filter using a circuit simulator due to the combination of the discrete time character of the filter and the functional test signals. Since the simulation is repeated for each fault and for various programmed settings, the required simulation time becomes unacceptable. By replacing the switched capacitors by equivalent models consisting of resistors and current sources, a time continuous equivalent is obtained, which allows for fast simulations. This equivalent model was used to analyze all faults in the decoding logic, operational amplifiers, resistive divider, etc. Faults affecting the clock signals of the switched capacitors and faults inside those capacitors were analyzed separately.

Analysis of 55 most likely faults identified 5 faults that were not detected by functional tests. Two of these could be detected by adding an additional functional test, but the remaining three faults were inherently undetectable. However, further analysis of these five faults revealed that they can be eliminated by minor changes in the routing of the layout.

7.9 Conclusion

Analog test issues are more qualitative than quantitative, therefore, digital DfT measures that address quantitative issues are not very successful with analog circuits. Owing to the non-binary nature of circuit operation, analog circuits are influenced by process defects in a different manner than digital circuits. Furthermore, in analog circuits, circuit design and layout techniques are utilized to maximize the performance. Therefore, these circuits show a greater sensitivity to parametric variations compared to their digital counterparts. Many manufacturing process defects that do not influence digital circuit performance, may affect analog circuits significantly. In general, subtle design and manufacturing process sensitivities give rise to many parametric failures. This requires a careful investigation of defects in analog circuits, modeling related aspects and detection strategies. In this chapter, a realistic process defect oriented simulation-before-test (SBT) testability methodology for analog circuits is proposed. A defect simulator, for sprinkling defects onto the layout, and a circuit simulator, for fault simulation, are the key components of the methodology. The circuits of moderate complexity can be analyzed. However, bigger circuits must be divided for the analysis.

A process defect based analog test approach contrasts with the specification based analog testing. Both approaches have their merits, However, there is a growing consensus that a synergy between the two will result in a better test quality as well test economics. The potential of IFA techniques is assessed on two standpoints: (a) contribution of IFA in testing silicon devices in a production environment and, (b) contribution of IFA in robust analog design against process defects, quantifying the fault coverage of analog tests, and examining the practicality of analog DfT schemes. In this chapter, we have presented results on both of these aspects of IFA.

To assess the contribution of IFA in solving analog test issues, a series of experiments were conducted on a Class AB amplifier in a production test environment. In these experiments, the effectiveness of IFA based test is compared to that of conventional specification based tests. IFA based tests for the Class AB amplifier were appended to the conventional test. Results of the exercise show that the vast majority of failures can be detected by simple IFA based test vectors. A fraction (0.4%) of the total tested devices found to be faulty by the conventional test, however, was not detected by the IFA based test. The subsequent analysis revealed that more than 85% of such escapes were due to un-modeled faults. IFA based tests are simpler compared to the conventional tests and, therefore, can be applied by inexpensive automatic test equipment (ATE) in a production test environment. Both of these aspects result in test cost reduction. The test cost reduction for the Class AB amplifier in the production environment is estimated to be 30%.

These experiments lead to some broad conclusions. Tighter the parametric specifications of an analog circuit, less effective an IFA based test is likely to be. This is because natural process variations with tighter parametric requirements will contribute to a larger number of device failures. These are un-modeled faults/failures, therefore, the effectiveness of IFA based test is lowered. Furthermore, the effectiveness of a process defects based yield loss model diminishes significantly with tightening of the parametric requirements. Hence, the application of IFA based tests to analog circuits with relatively relaxed specifications and higher functional complexity is likely to be more successful. Better control of the manufacturing process (high Cp and Cpk) should also have positive influence on the effectiveness of IFA based test. A limited functional test together with an IFA based test should be another way to avoid such escapes. Thus, strengths of both test methods can be exploited for economic and quality gains.

Assessment of IFA in an analog design environment was carried out using a flash A/D converter. Fault coverage for analog circuits is often not determined. In the absence of fault coverage numbers, effectiveness of any DfT measure cannot be quantified. It is demonstrated that the IFA technique can be exploited to fault grade the flash A/D converter for existing (specification) test practices. The results of the analysis showed that the fault coverage of the specification test was relatively high on the data path. However, it was relatively poor on the control path (clock, bias generators). Almost all specification based tests are targeted towards the faithfulness of the data path and the control path is assumed to be tested implicitly. Separate DfT solutions were proposed for clock and bias generators to improve the fault coverage and to simplify the test.

IFA is has limitations. First and foremost is CPU intensiveness of the method. Application of the analysis requires CAD tools. Circuits of moderate transistor count complexities can only be analyzed by a CAD tool. The circuits with higher complexities must be partitioned or modeled at higher abstraction levels for the analysis.

References

1. B. Atzema, E. Bruls, M. Sachdev and T. Zwemstra, "Computer-Aided Testability Analysis for Analog Circuits," *Proceedings of the Workshop on Advances in Analog Circuit Design*, April 1995.

2. R. Becker, et al., "PACT - A Programmable Analog CMOS Transmission Circuit for Electronic Telephone Sets," Proceedings of European Solid State Circuits Conference, 1993, pp. 166-169.

3. F.P.M. Beenker, "Testability Concepts for Digital ICs," *Ph.D. Thesis, University of Twente, Netherlands,* 1994.

4. R.S. Berkowitz, "Conditions for Network-element-value Solvability," *IRE Transactions on Circuit Theory*, vol. CT-9, pp. 24-29, March 1962.

5. E.M.J.G. Bruls, "Reliability Aspects of Defects Analysis," *Proceedings of European Test Conference*,1993, pp. 17-26.

6. E. Bruls, "Variable supply voltage testing for analogue CMOS and bipolar circuits," *Proceedings of International Test Conference,* 1994, pp. 562-571.

7. P. Duhamel and J.C. Rault, "Automatic Test generation Techniques for Analog Circuits and Systems: A Review," *IEEE Transactions on Circuits and Systems*, vol. CAS-26, no. 7, pp. 411-440, July 1979.

8. N.J. Elias, "The Application of Statistical Simulation to Automated the Analog Test Development," *IEEE Transactions on Circuits and Systems*, vol. CAS-26, no. 7, pp. 513-517, July 1979.

9. B.R. Epstein, M. Czigler and S.R. Miller, "Fault Detection and Classification in Linear Integrated Circuits: An Application of Discrimination Analysis and Hypothesis Testing," *IEEE Transactions on Computer Aided Design of Integrated Circuits and Systems*, vol. 12, no. 1, pp. 102-112, January 1993.

10. R.J.A. Harvey, A.M.D. Richardson, E.M.J. Bruls and K. Baker, "Analogue Fault Simulation Based on Layout Dependent Fault Models," *Proceedings of International Test Conference*, 1994, pp. 641-649.

11. C.F. Hawkins and J.M. Soden, "Electrical Characteristics and Testing Considerations for Gate Oxide Shorts in CMOS ICs", *Proceeding of International Test Conference*, 1985, pp. 544-555.

12. G.J. Hemink, B.W. Meijer and H.G. Kerkhoff, "TASTE: A Tool for Analog System Testability Evaluation," *Proceeding of International Test Conference*, 1988, pp. 829-838.

13. W. Hochwald and J.D. Bastian, "A DC Approach for Analog Dictionary Determination," *IEEE Transactions on Circuits and Systems*, vol. CAS-26, no. 7, pp. 523-529, July 1979.

14. H.H. Huston and C.P.Clarke, "Reliability Defect Detection and Screening During Processing - Theory and Implementation," *Proceedings of International Reliability Physics Symposium,* 1992, pp. 268-275.

15. A.T. Johnson, Jr., "Efficient Fault Analysis in Linear Analog Circuits," *IEEE Transactions on Circuits and Systems*, vol. CAS-26, no. 7, pp. 475-484, July 1979.

16. F.C.M. Kuijstermans, M. Sachdev and L. Thijssen, "Defect Oriented Test Methodology for Complex Mixed-Signal Circuits," *Proceedings of European Design and Test Conference*, 1995, pp. 18-23.

17. M. Mahoney, "*DSP-Based Testing of Analog and Mixed-Signal Circuits*", Los Alamitos, California: IEEE Computer Society Press, 1987.

18. W. Maly, F.J. Ferguson and J.P. Shen, "Systematic Characterization of Physical Defects for Fault Analysis of MOS IC Cells," *Proceeding of International Test Conference*, 1984, pp. 390-399.

19. W. Maly, W.R. Moore and A.J. Strojwas, "Yield Loss Mechanisms and Defect Tolerance," *SRC-CMU Research Center for Computer Aided Design, Dept. of Electrical and Computer Engineering, Carnegie Mellon University, Pittsburgh, PA 15213.*

20. W. Maly, A.J. Strojwas and S.W. Director, "VLSI Yield Prediction and Estimation: A Unified Framework," *IEEE Transactions on Computer Aided Design*, vol. CAD-5, no.1, pp 114-130, January 1986.

21. W. Maly, "Realistic Fault Modeling for VLSI Testing," *24th ACM/IEEE Design Automation Conference*,1987, pp.173-180.

22. A. Meixner and W. Maly, "Fault Modeling for the Testing of Mixed Integrated Circuits," *Proceeding of International Test Conference*, 1991, pp. 564-572.

23. R. Mehtani, B. Atzema, M. De Jonghe, R. Morren, G. Seuren and T. Zwemstra,"Mix Test: A Mixed-Signal Extension to a Digital Test System," *Proceedings of International Test Conference*, 1993, pp. 945-953.

24. L. Milor and V. Visvanathan, "Detection of Catastrophic Faults in Analog Integrated Circuits," *IEEE Transaction on Computer Aided Design of Integrated Circuits and Systems*, vol. 8, pp. 114-130, February 1989.

25. L. Milor and A. Sangiovanni-Vincentelli, "Optimal Test Set Design For Analog Circuits", *International Conference on Computer Aided Design*, 1990, pp. 294-297.

26. N. Navid and A.N. Willson, Jr., "A Theory and an Algorithm for Analog Fault Diagnosis," *IEEE Transactions on Circuits and Systems*, vol. CAS-26, no. 7, pp. 440-456, July 1979.

27. A. Pahwa and R. Rohrer, "Band Faults: Efficient Approximations to Fault Bands for the Simulation Before Fault Diagnosis of Linear Circuits," *IEEE Transactions on Circuits and Systems,* vol. CAS-29, no. 2, pp. 81-88, February 1982.

28. M.J.M. Pelgrom and A.C. van Rens, "A 25 Ms/s 8-Bit CMOS ADC for Embedded Applications," *Proceedings 19th of European Solid State Circuits Conference*, 1993, pp. 13-16.

29. R.J. van de Plassche, "*Integrated Analog-to-Digital and Digital-to-Analog Converters*," Dordrect: Kluwer Academic Publishers. 1994.

30. R.W. Priester and J.B. Clary, "New Measures of Testability and Test Complexity for Linear Analog Failure Analysis," *IEEE Transactions on Circuits and Systems*, vol. cas-28, no.11, pp. 1088-1092, November 1981.

31. L. Rapisarda and R.A. Decarlo, "Analog Multifrequency Fault Diagnosis," *IEEE Transactions on Circuits and Systems*, vol. cas-30, no.4, pp. 223-234, April 1983.

32. R. Rodriguez-Montanes, E.M.J.G. Bruls and J. Figueras, "Bridging Defects Resistance Measurements in CMOS Process," *Proceeding of International Test Conference*, 1992, pp. 892-899.

33. M. Sachdev, "Catastrophic Defect Oriented Testability Analysis of a Class AB Amplifier," *Proceedings of Defect and Fault Tolerance in VLSI Systems*, October 1993, pp. 319-326.

34. M. Sachdev, "Defect Oriented Analog Testing: Strengths and Weaknesses," *Proceedings of 20th European Solid State Circuits Conference*, 1994, pp. 224-227.

35. M. Sachdev, "A Defect Oriented Testability Methodology for Analog Circuits," *Journal of Electronic Testing: Theory and Applications*, vol. 6, no. 3, pp. 265-276, June 1995.

36. M. Sachdev and B. Atzema, "Industrial Relevence of Analog IFA: A Fact or A Fiction," *Proceedings of International Test Conference*, 1995, pp. 61-70.

37. M. Sachdev, "A DfT Method for Testing Internal and External Signals in A/D Converters", *European patent application no.* 96202881.7, 1996.

38. R. Saeks, A. Sangiovanni-Vincentelli and V. Vishvanathan, "Diagnosability of Nonlinear Circuits and Systems--Part II: Dynamical case," *IEEE Transactions on Circuits and Systems*, vol. cas-28, no.11, pp. 1103-1108, November 1981.

39. A.E. Salama, J.A. Starzyk and J.W. Bandler, "A Unified Decomposition Approach for Fault Location in Large Analog Circuits," *IEEE Transactions on Circuits and Systems*, vol. cas-31, no.7, pp. 609-622, July 1984.

40. J.P. Shen, W. Maly and F.J. Ferguson, "Inductive Fault Analysis of MOS Integrated Circuits," *IEEE Design and Test of Computers,* vol. 2, no. 6, pp. 13-26, 1985.

41. M. Slamani and B. Kaminska, "Analog Circuit Fault Diagnosis Based on Sensitivity Computation and Functional Testing," *IEEE Design & Test of Computers*, vol. 9, pp. 30-39, March 1992.

42. J.M. Soden and C.F. Hawkins, "Test Considerations for Gate Oxide Shorts in CMOS ICs," *IEEE Design & Test of Computers*, vol. 2, pp. 56-64, August 1986.

43. J.M. Soden and C.F. Hawkins, "Electrical Properties and Detection Methods for CMOS IC Defects," *Proceedings of European Test Conference*, 1989, pp. 159-167.

44. M. Soma, "Fault Modeling and Test Generation for Sample and Hold Circuits," *Proceedings of International Symposium on Circuits and Systems*, 1991, pp. 2072-2075.

45. M. Soma, "An Experimental Approach to Analog Fault Models," *Proceedings of Custom Integrated Circuits Conference,* 1991, pp. 13.6.1-13.6.4.

46. M. Soma, "A Design for Test Methodology for Active Analog Filters," *Proceedings of International Test Conference*, 1990, pp. 183-192.

47. T.M. Souders and G.N. Stenbakken, "A Comprehensive Approach for Modeling and Testing Analog and Mixed Signal Devices," *Proceeding of International Test Conference*, 1990, pp. 169-176.

48. H. Sriyananda and D.R. Towill, "Fault diagnosis Using Time-Domain Measurements," *Proceedings of Radio and Electronic Engineer*, vol.9, no. 43, pp. 523-533, September 1973.

49. M. Syrzycki, "Modeling of Spot Defects in MOS Transistors," *Proceedings International Test Conference,* 1987, pp. 148-157.

50. T.N. Trick, W. Mayeda and A.A. Sakla, "Calculation of Parameter Values from Node Voltage Measurements," *IEEE Transactions on Circuits and Systems,* vol. CAS-26, no. 7, pp. 466-474, July 1979.

51. V. Vishvanathan and A. Sangiovanni-Vincentelli, "Diagnosability of Nonlinear Circuits and Systems--Part I: The DC case," *IEEE Transactions on Circuits and Systems*, vol. CAS-28, no.11, pp. 1093-1102, November 1981.

52. K.D. Wagner and T.W. Williams, "Design for Testability of Mixed Signal Integrated Circuits," *Proceeding of International Test Conference*, 1988, pp. 823-828.

53. A. Walker, W.E. Alexander and P. Lala, "Fault Diagnosis in Analog Circuits Using Elemental Modulation," *IEEE Design & Test of Computers*, vol. 9, pp. 19-29, March 1992.

54. H. Walker and S.W. Director, "VLASIC: A Catastrophic Fault Yield Simulator for Integrated Circuits," *IEEE Transactions on Computer Aided Design of Integrated Circuits and Systems*, vol. CAD-5, pp. 541-556, October 1986.

55. R.H.Williams and C.F. Hawkins, "Errors in Testing," *Proceeding of International Test Conference*, 1990, pp. 1018-1027.

CHAPTER 8

Conclusion

In concluding this book we summarize the accomplishments. We began with emphasizing the relevance of testing in general and the structural testing in particular. The main contributions of the defect oriented testing are summarized and at the same time its limitations are also highlighted. Furthermore, future trends and research directions are recommended.

8.1 The Test Complexity

Imperfections in the manufacturing process necessitate testing of the manufactured ICs. The fundamental objective of testing is to distinguish between good and faulty ICs. This objective can be achieved in several ways. Earlier, when ICs were relatively less complex, this objective was achieved by the functional testing. However, as the complexity of the fabricated ICs increased, it was soon discovered that the application of the functional test is rather expensive in test resources and is inefficient in catching the manufacturing process imperfections. The exponential increase in the cost of functional testing led to tests that are not functional in nature, but are aimed to detect the possible faulty conditions in ICs. The circuit is analyzed for faulty conditions and tests are generated to test for such conditions. Like any other analysis, this fault analysis also requires a model (or abstraction) to represent likely faults in ICs with an acceptable level of accuracy. This type of fault model based testing is known as structural testing. The name structural test comes from two counts. First, the testing is carried out to validate the structural composition of the processed design rather than the its function and, second, the test methodology has a structural basis, i.e., the fault model for test generation.

Structural testing gained popularity in the 70s and the 80s when LSI complexity forced researchers to pay attention to test cost reduction [12]. Structural DfT methodologies like scan path, level sensitive scan design (LSSD) emerged for digital circuits [3,4,13]. These DfT methods became popular because their application could change distributed sequential logic into a big unified shift-register for testing purposes. As a result, the overall test complexity is reduced. Owing to these techniques, the test generation and the fault grading for complex digital circuits became possible.

At the same time, it was discovered that classical voltage based test methods for digital CMOS ICs are grossly inadequate in ensuring the desired quality and reliability levels. Many commonly occurring defects like gate oxide defects often are not detected by logic tests. Therefore, such escaped defects are quality and reliability hazards. This increased quality awareness brought in new test techniques like quiescent current measurements (QCM), or I_{DDQ} test as it is popularly known, in the test flow for digital CMOS ICs [1,5]. Arguably I_{DDQ} is an effective test method for catching manufacturing process defects. Nevertheless, there is consensus among test professionals that both voltage as well as I_{DDQ} based tests are needed for ensuring the quality of ICs.

8.2 Defect Oriented Testing

The circuit layout influences the impact of a defect, and thus the faulty circuit behavior to a large extent. This information is often ignored while developing fault models at transistor or logic level. A defect oriented test method takes into account the sensitivity of a given layout and defect data from the targeted fabrication plant to generate realistic or likely defects in the given circuit.

Once, the information about the likely defects in a given circuit is known, tests can be generated for high defect (fault) coverage. Alternatively, remedial actions in design and/or test may be taken to improve the quality and economics of the test. The same information may be exploited to improve the yield of an IC.

In principle, the defect oriented test method is independent of technology and design style. The method may be used for digital, analog or mixed signal circuits in purely CMOS, BiCMOS or bipolar technologies. In spite of the method's wide applicability, we restricted ourselves to the CMOS environment because of the overwhelming benefits and popularity of the CMOS technology. Furthermore, the level of information about defects that is available for CMOS technology is unmatched in other technologies. Finally, the application of a key CAD tool, VLASIC, is restricted to CMOS circuits. Considering these practical factors, we took representative examples of CMOS circuits to analyze through the defect oriented test method. These circuits include purely digital circuits (flip-flops, scan chains), quasi-digital/analog circuits (DRAM, SRAM) and analog circuits (Class AB amplifier, A/D converter).

8.2.1 Strengths of Defect Oriented Testing

Defect information can be exploited in many ways during the development of an IC. The defect oriented testing is receiving substantial attention in the industry. It has been applied to a complex, mixed-signal and large volume single chip TV IC at Philips Semiconductors to improve the yield and quality [7]. Similarly, other companies and institutions are paying attention to defect oriented testing. The salient advantages of this method are as follows:

• **Shorter and efficient production tests**
    The application of defect oriented method often results in shorter and effective production tests. Tests are directed towards a particular class of defects likely to occur in the production environment and cause yield loss in a particular circuit type. For example, the work of Dekker et al. [2] on testing

of SRAMs and that of Sachdev [8] on testing of class AB amplifier illustrates the potential of the method.

• **Improved and robust design**

The information on what can go wrong in a basic cell can be exploited to improve the design so that the detection of difficult to test defects is simplified. For example, a set of testable flip-flop configurations were devised. Besides providing testability benefits, these flip-flop configurations can withstand higher clock-skew. Furthermore, proposed configurations have approximately 25% less setup and hold time, and have better metastability than the conventional configurations. Alternatively, the defect information may be exploited to design robust or defect insensitive circuits and layouts.

• **Defect based DfT techniques**

The defect information can also be exploited to devise innovative test modes such that defects are quickly and easily tested. For example, scan chain transparency results in testing of defects in scan chains quickly and efficiently. Similarly, an I_{DDQ} based parallel RAM test methodology results in efficient detection of defects. In analog circuits also it is possible to devise DfT techniques to test hard-to-detect defects. Furthermore, one can fault grade analog tests for realistic defects.

8.2.2 Limitations of Defect Oriented Testing

Every test method has its constraints and limitations. The defect oriented test method is no exception. Owing to CAD tool limitations, the application of this method is limited to relatively small macros and building blocks. At present, research effort is directed towards relaxing such constraints. Relatively large number of defects should be sprinkled onto the analyzed layout to generate a realistic fault list. As a result large number of simulations must be performed to ascertain the effectiveness of the given set of tests. For accurate results, these simulations need to be performed at the Spice level requiring large amount of CPU resources. Alternatively, a high level model requires substantially lower effort. Present research efforts at Philips Research Labs and elsewhere continue to automate most the mundane work and, hence, speed up the analyses.

8.3 Future Directions

It is always difficult to predict the future. It is best to look backwards to predict the future trends. Testing has come a long way. In the early days of semiconductor industry, testing was merely a verification of functionality. Quality and economics issues were non-existent. In late 70s and 80s, growing semiconductor industry understood the futility of functional testing and started to pay attention to fault modeling, structural testing, DfT techniques, etc. Alternative test techniques, like I_{DDQ}, emerged as "quality" supplements. At the same time, quality and economics became core business issues. With the increasing competition, it was no longer sufficient to be able to produce a product. The focus shifted on how efficiently and economically one can make the product. Testing is recognized as the bottleneck and the last checkpoint for ensuring product quality and reliability. As a result, in the first half of 90s, we witnessed a number of studies reporting benefits of incorporating non-Boolean (I_{DDQ}) tests in the test suite.

What next? ICs are going to be more complex and faster. For example, Intel projects a 1 GHz processor with 100 million transistors by the turn of the century [9]. Similarly, 256 Mbit DRAMs are being designed with 0.2μ design rules and 1-Gbit DRAMs are expected to appear in the later half of 1990s [11]. It is clear that the integration capability has grown much faster than the Moore's law and this trend is likely to continue in the near future [10,14]. In spite of these advances testing is a bottleneck and is expected to stay as the bottleneck. At the same time, the deep sub-micron test issues (reduced power supply voltage, reduced noise margin and increased transistor sub-threshold current) will gain relevancy and force researchers to look for innovative test solutions. Recently, it was reported [6] that there are some fault classes that cannot be modeled as a 2-dimensional disc shaped defect and require a 3-dimensional contaminations model. More research is needed in this area to improve the quality of fault models.

References

1. S. Chakravarty and P.J. Thadikaran, *Introduction to I_{DDQ} Testing*, Boston: Kluwer Academic Publishers, 1997.

2. R. Dekker, F. Beenker and L. Thijssen, "Fault Modeling and Test Algorithm Development for Static Random Access Memories," *Proceedings of International Test Conference*, 1988, pp. 343-352.

3. E.B. Eichelberger and T.W. Williams, "A Logic Design Structure for LSI Testability," *Journal of Design Automation and Fault Tolerant Computing*, vol. 2, no. 2, pp. 165-178, May 1978.

4. S. Funatsu, N. Wakatsuki and T. Arima, "Test Generation Systems in Japan," *Proceedings of 12th Design Automation Symposium*, 1975, pp. 114-122.

5. R.K. Gulati and C.F. Hawkins, *I_{DDQ} Testing of VLSI Circuits*, Boston: Kluwer Academic Publishers, 1993.

6. J. Khare and W. Maly, "Inductive Contamination Analysis (ICA) with SRAM Applications," *Proceedings of International Test Conference*, 1995, pp. 552-560.

7. L. Nederlof, "One Chip TV," *Proceedings of International Solid State Circuits Conference*, 1996, pp. 26-29.

8. M. Sachdev, "A Defect Oriented Testability Methodology for Analog Circuits," *Journal of Electronic Testing: Theory and Applications*, vol. 6, no. 3, pp. 265-276, June 1995.

9. J. Schutz, "The Evolution of Microprocessor Design in Response to Silicon Process Evolution," *Proceedings of 21st European Solid State Circuits Conference*, 1995, pp. 16-19.

10. P. Singer, "1995: Looking Down the Road to Quarter-Micron Production," *Semiconductor International*, vol. 18, no. 1, pp. 46-52, January 1995.

11. E. Takeda et al, "VLSI Reliability Challenges: From Device Physics to Wafer Scale Systems," *Proceedings of IEEE*, vol. 81, no. 5, pp. 653-674, May 1993.

12. P. Varma, A.P. Ambler and K. Baker, "An Analysis of the Economics of Self-Test," *Proceedings of International Test Conference*, 1984, pp. 20-30.

13. T.W. Williams and K.P. Parker, "Design for Testability--A Survey," *Proceedings of the IEEE*, vol. 71, no. 1, pp. 98-113, January 1983.

14. -- *"The National Technology Roadmap for Semiconductors,"* San Jose: Semiconductor Industry Assciation, 1995.

Index